Telecommunications:
Concepts, Development, and Management

5 B9S37.35

MARLEN LEDBETTER
EE 5377/SMU
SUMMER/89

Telecommunications:
Concepts, Development, and Management

W. John Blyth
MCI Telecommunications Corporation

Mary M. Blyth
Detroit College of Business

Glencoe Publishing Company
Mission Hills, California

Developmental editor: Dennis L. Gladhill
Acquisitions editor: Bobby L. Yount
Project coordinator: John Gastineau
Editing and production services: The Chestnut House Group, Inc.

Send all inquiries to:
Glencoe Publishing Company
15319 Chatsworth Street
Mission Hills, California 91345

Printed in the United States of America

Library of Congress Cataloging in Publication Data

Blyth, W. John.
Telecommunications: concepts, development, and management.

Includes bibliographies and index.
1. Telecommunication—United States. 2. Telecommunication—Management.
I. Blyth, Mary M. II. Title.
HE7775.B53 1984 384 84-12392

ISBN 0-02-680840-4

4 5 6 90 89 88

To David and Dennis

Contents

Preface

A new era in telecommunications—the era of deregulation of the industry and divestiture of AT&T—began in 1984. In this new environment, which offers exciting opportunities for enhanced services and reduced costs, there is a sudden, critical need for telecommunications management that is effective by its structured, scientific approach.

This book is a first step toward meeting that need—a college-level textbook on the fundamentals of telecommunications. It provides an overview of the development and present structure of the telecommunications industry, the fundamental concepts of telecommunications, and the basic skills required for cost-effective telecommunications management.

Because *Telecommunications: Concepts, Development, and Management* is written in a nontechnical style, it will be useful to persons without previous training in telephony, electronics, or data processing. Since each chapter is relatively independent of the other chapters, this book may also be used to teach specific topics in telecommunications in business courses or industry seminars.

Professionals who already have specific knowledge in a particular aspect of telecommunications will find this text valuable in broadening their knowledge of the field.

The text includes the following instructional features:

- ☐ summaries at the end of each chapter to review important points
- ☐ review questions at the end of each chapter to enable readers to evaluate their comprehension of the material
- ☐ a references and bibliography section at the end of each chapter to enable the reader to explore the topics in greater depth
- ☐ a list of professional associations
- ☐ a list of organizations that sponsor seminars

- ☐ a list of telecommunications equipment suppliers
- ☐ samples of forms and financial research documents used to justify new telecommunication systems
- ☐ a glossary of terms
- ☐ an index

An instructor's guide containing chapter objectives, chapter outlines/lecture notes, answers to review questions, and examination questions for each chapter is available from the publisher.

The text was field tested prior to publication by the authors at Oakland Community College in Auburn Hills, Michigan, and by Dr. Thomas Muth at Michigan State University in East Lansing, Michigan.

Welcome to the exciting new world of telecommunications! It is potentially one of the most important cost-saving areas of business operation in the twentieth century. This book is designed to be your passport into this dynamic field.

W. John Blyth
Mary M. Blyth

Acknowledgments

We wish to express our appreciation to the following telecommunications professionals for their assistance in the development of this text: Daniel Grove, Motorola Corporation; Edna F. Afiriyie, American Telephone and Telegraph Company; Marvin G. Pridgeon, Ameritech Corporation; Ernest O. Miller, Michigan Bell Telephone Company; and John Gorski, Ben Berman, and David Shaffer, MCI Telecommunications Corporation.

We thank Mary McIntyre, Director of Libraries, Detroit College of Business; Gerald R. Lamphere, Chairperson of Data Processing, Detroit College of Business; and Thomas Muth, Michigan State University, for their invaluable assistance.

Our thanks also go to the following people, who read and commented on the manuscript in its various stages: Marcia Anderson, Joyce Arntson, John L. Fike, Marie E. Flatley, Arthur Goodall, Ron Kapper, Larry Smith, Kathleen Wagoner, and Jim Wildhaber.

Telecommunications:
Concepts, Development, and Management

Introduction to Telecommunications 1

We are all familiar with the telephone. We know that it permits us to communicate with another person nearly anywhere in the world. We know this because we have had firsthand experience with the telephone. When we use the word *telephone,* we can be quite sure that other people know exactly what we mean.

This is not the case, however, with telecommunications. *Telecommunications* is a relatively new word in our vocabularies. In fact, while the word *telephone* has been in existence for over a century, *telecommunications* was not even listed in dictionaries until the mid-1960s.

The Field of Telecommunications

To some people, telecommunications is synonymous with voice communications; to others it is synonymous with data communications, broadcasting, or any of the other electronic components of the entertainment industry. While each of these areas is an important part of telecommunications, it is only a part. The field of telecommunications is a broad one that includes voice communications, data communications, image communications, and message communications.

The Term Defined

The prefix *tele* is derived from the Greek root meaning "distant" or "at a distance." *Telecommunication* is the process of transmitting information over a distance by electrical or electromagnetic systems. Information may take the form of voice, data, image, or messages. Electromagnetic transmission systems include telephone lines, cables, microwaves, satellites, and light beams.

Telecommunication Systems

Two characteristics identify a telecommunication system:

1. communication over a distance
2. transmission by electromagnetic means

Four principal types of telecommunication systems are available today: voice, data, message, and image.

Voice communication is known as *telephony*. Voice communication systems transmit spoken words over telephone networks in the form of electrical energy that varies in amplitude with the sound variations being transmitted. These systems include public and private, local and long distance services.

Data communication systems are networks of components and devices organized to transmit data from one location to another—usually from one computer or computer terminal to another. The data is transmitted in coded form over electrical transmission facilities. Data transmission systems aim to provide faster information flow by reducing the time spent in collecting and distributing data.

Similar to data communication systems are message and image systems. Message systems, such as TWX and Telex, send messages in data form; i.e., telegram or teletypewriter messages. They are *message systems* because they are used to transmit messages rather than conversation. Message systems provide a faster alternative to the postal service. *Image* or *facsimile systems* send exact messages—pictures, sketches, diagrams, and graphs—in data form. Data systems, message systems, and image systems all transmit intelligence in coded form by means of a stream of pulses that represent digital signals.

Computers and Communications

Communications technology and computer technology developed along parallel lines. Early telephone systems utilized electromechanical switches to connect calls between telephones. Similarly, early computers used electromechanical equipment controlled by punched cards to perform sorting and computing operations.

Both industries contributed to the development of solid-state electronics and use it in their modern systems. Thus, the technology of these two industries is compatible. Combining a number of data processing units using the telecommunication network, each unit performs a particular function, and the system can operate at maximum efficiency. Today's telephones are capable of providing much more than plain old telephone service (POTS). Computerized components in the telephone system control several functions such as call forwarding, call waiting, and automatic dialing. Electronic telephone sets, described as "smart" telephones, perform many functions within the telephone itself rather than in the switching system. Combining features of data processing and communications greatly enhances the capabilities of both systems.

"The convergence of communications and computers is natural, since both deal with information. Computers store and manipulate information; communications systems transmit the information from one point to another," observes Edward Cornish, editor of *The Futurist*.[1]

Telematics, the marriage of telecommunications and computer technology, has made the information age possible, and with it exciting opportunities for enhanced productivity and profits. The word *telematics* is derived from *telematique*, a term coined by French scientists Simon Nora and Alain Minc to describe the merging of telecommunications with computers and television. It was translated into English as telematics. Professor Anthony Oettinger of Harvard University coined the term *compunications* to describe the same union; however, this term has not gained widespread acceptance. The marriage has produced a number of offspring, including teleconferencing, telemarketing, telecommuting, telecourse, telejournal, telemedicine, and teleflora.

Telecommunications evolved as a branch of electronics, specifically, electrical engineering. As with most disciplines, electronics has its own special language. The technical aspects of the language, oriented toward electrical circuitry, often deter laypeople from studying telecommunications. However, the increasing use of electronic information handling has made it necessary for businesspersons—and many others as well—to understand the terminology and underlying principles of telecommunications.

Recognizing the need for people to be familiar with computers, many schools and colleges are requiring students to develop computer literacy. Telecommunications is no less important. As we progress toward "the paperless office," "the cashless society," and "the home office," persons from all walks of life are finding that a knowledge of the language and concepts of telecommunications is becoming a must. The intention of this textbook is to describe the basic terminology and concepts of telecommunications in nontechnical language so that persons with no special training or background in the field can understand them.

Telecommunications in Modern Society

The United States is rapidly becoming an information-based society, and telecommunications is central to this development. The availability of nearly "instant information" made possible by

1. Edward Cornish, "The Coming of an Information Society," *Communications Tomorrow: The Coming of the Information Society*, selections from *The Futurist* (Bethesda, Md.: World Future Society, 1982), p. 45.

telecommunications technology is changing our jobs, our business organizations, and our personal lives.

Telecommunications in Business Organizations

Information is the lifeline of modern business; it provides the basis for all business activities. Managers apply judgment to the information available to make their decisions—decisions that have a strong impact upon the success of any enterprise. Recognizing their dependency upon information, businesses have begun to focus on the management of this vital resource.

The present trend in information management is to establish formal *management information systems (MIS)* to make information immediately available to various levels of management throughout an organization. Some companies have created separate Management Information System Departments to coordinate the flow of information throughout the organization. Telecommunications plays a key role in these departments.

Telecommunications makes possible the technique of *distributed processing,* which links intelligent terminals to central computer processing facilities through the use of communication lines. Telecommunications also makes possible the instant availability of information from a centralized data base (facts arranged in computer files for access and retrieval) within an organization. With a data base, information can be shared among the various departments of an organization and among remotely located branches. Thus, many different people with different job responsibilities can access the data base from terminals, often referred to as *workstations,* located in their work areas.

Some companies have established industry-wide data bases that can be accessed from virtually anywhere via telecommunication lines. Companies that maintain data base services offer access to their data base for a subscription fee. These information centers are available for many industries and professions, including transportation, banking, investments, publications, law, and medicine. The data bases are protected from unauthorized access by security measures that require user identification via a password or series of passwords.

WESTLAW, one commercial data base system, was created by West Publishing Company, a publisher that has served the legal profession for over a century. The WESTLAW system connects remote terminals located in law offices, courts, and government agencies with a centralized data base by means of telephone lines. Users enter commands or inquiries at a typewriter-like keyboard; responses are displayed on the cathode ray tube (CRT), or terminal screen. Users can record information appearing on the screen on paper by means of an associated printer. The system is protected from unauthorized access by a multistep sign-on procedure.

This system offers legal research capabilities that permit a lawyer to scan large numbers of cases and identify relevant ones in a few seconds. The system has two basic advantages:

1. speed in finding the research material
2. search capabilities not possible using books

The system's unique search capabilities allow users to access case summaries using terms other than those indexed in law books. For example, users can locate a case by entering almost any term associated with that court decision—names of judges, witnesses, companies, and unusual nonlegal terms.

WESTLAW is constantly being updated; as new cases are reported, they are added to the data base. The system has improved legal research by allowing users to accomplish more research in less time and by providing new search capabilities.

Industries That Use Telecommunications

Many organizations are using telecommunications effectively to enhance their services or to provide new services. Additionally, there are industries whose very existence depends upon telecommunications. Among these are airlines, banking, investments, credit card operations, and hotel/motel reservation systems.

Airlines All major airlines in the United States have computerized systems for handling reservations. To make a reservation, a person dials a local telephone number that connects to a regional reservation center, probably located in a distant city. The regional center is linked to a national center by telephone lines. The reservation clerk keys in the desired destination on a remote terminal to access the central computer and obtain information about possible flights. Information displayed on the terminal screen lists the flights on which seats are available and the fares. When the traveler selects a flight, the clerk enters the details of the booking into the system, and the computer's seat inventory is updated. The cancellation of reservations also is handled automatically; the computer cancels the reservation and adds to the number of seats available on the flight. In addition, the computer keeps a waiting list of passengers desiring reservations and their telephone numbers. Reservation systems are interlinked with those of other airlines so that connecting flights can be booked.

Telecommunications also permits computer operators to monitor air traffic from control towers. A computer system identifies each approaching plane and tracks its altitude and speed, enabling air traffic controllers to give directions for landing.

A number of other industries whose business depends heavily on advance reservations employ telecommunication systems similar to those used by the airlines. These industries include hotels and motels, car rental agencies, ship lines, and railroads.

Banking Another major industry dependent upon telecommunications technology is banking. Banks use remote terminals and telecommunication lines to update customers' accounts. Tellers located either at main offices or branch offices insert the customer's passbook into a computer terminal and key in relevant information; the terminal prints the entry in the passbook and transmits the information to update the central computer files.

In some cases, customers can use their pushbutton telephones to perform certain banking transactions such as paying bills, transferring funds, and determining their bank balances. The telephone functions as an input terminal when it is connected to the bank's central computer by telephone lines. The user communicates with the computer by entering appropriate codes on the telephone keypad.

Banks also use telecommunication lines to provide automatic teller service. Automatic teller terminals are connected to the bank's central computer via communication lines. To process a transaction, the customer inserts an identifying bank card into the terminal and keys in a personal identification number. After the computer performs appropriate checks on the customer's identity and account status, the customer keys in a transaction code, and the machine completes the transaction. Automatic tellers are located in shopping malls, supermarkets, and places of employment as well as on bank premises. Many of these terminals accommodate bank cards issued by cooperating banks, credit unions, and savings and loan associations. Automatic tellers offer the convenience of 24-hour operation for the most frequently used services such as deposits, withdrawals, and payments to utilities and credit card accounts.

Credit Card Service The widespread use of credit cards has resulted in the creation of a new industry to serve as a clearinghouse for credit card transactions. A major function of this service is verifying accounts to reduce fraudulent use of credit cards. When a customer presents a card to be used for payment, the merchant calls the service company to obtain a credit status report. The service company searches its computer files and reports the status of the account. Account verification is a high-speed transaction; the entire process is completed in a few seconds. The service depends upon high-speed telecommunication between a centralized data base and merchants located virtually anywhere in the world.

Insurance Telecommunications enables insurance companies to operate more efficiently and to offer better service to their customers. Insurance companies with millions of policyholders generally have a home office and a number of branch offices located throughout the country. Policy records are kept in a central computer located in the home office, and the branch offices are connected to the home office computer via telecommunication lines. Each branch office is equipped with a terminal that permits it to communicate with the headquarters computer. When a policyholder requests information regarding coverage, a clerk in the branch office enters the request on the terminal keyboard. The computer accesses the files and reports the requested information on the terminal screen at the branch office. The entire process takes only a few seconds; thus, the clerk is able to respond to the customer's request without delay.

Investments Another industry that depends on the immediate communication of information is the investment business. Both the New York Stock Exchange and the American Stock Exchange—the two major stock exchanges in the country—are located in New York City. All trading of stocks they list is conducted on the floor of the exchanges. Stockbrokers throughout the country transmit their buy and sell orders over communication lines to their representatives at the stock exchange. As each transaction is completed, it is recorded on a computer and simultaneously transmitted to brokerage offices all over the country. Some brokerage offices display this information on a continuous tape, thus providing up-to-the-minute details of all trading transactions. Many brokerage offices also have display terminals connected to the New York computer. Stockbrokers use these terminals to obtain current price information on any stock from the computer. The brokerage industry relies heavily on rapid communication, since investors' trading decisions are based on the latest market prices.

Telecommunications in Personal Communications

Telecommunications has also enhanced personal communications. As our society has become increasingly mobile, telephone services have assumed greater importance in maintaining family and social ties. Long distance telephone calls have virtually replaced letter writing as a means of keeping in touch.

The telephone makes it possible to obtain information, to transact business without leaving home, to summon assistance in times of trouble, and to maintain social contacts.

New service features made possible by sophisticated telecommunications technology have also been favorably received. Many residential customers now enjoy the convenience of such computerized features as call waiting, automatic dialing, add-on, call fowarding, and cordless telephones.

The Restructuring of the Telecommunications Industry

In the United States, the restructuring of the telecommunications industry has taken place gradually over the last two decades; however, the process accelerated rapidly in the early 1980s. It involved the introduction and growth of competition, the trend toward deregulation, and the divestiture of the American Telephone and Telegraph Company (AT&T).

The telecommunications industry in the United States is a private enterprise operating as a regulated monopoly. The Federal Communications Commission (FCC), which was created by the Communications Act of 1934, is charged with regulating all interstate and foreign telephone, telegraph, and broadcast communications. The FCC sets public policy and supervises the utilities in the execution of that policy. The Communications Act of 1934 stipulated that, as much as possible, the telephone industry should be operated so telephone service is available to all the people of the United States at reasonable charges.

For half a century, the FCC maintained that the telecommunications industry was a "natural monopoly" and that no geographic location should be served by more than one telephone company. This policy was designed to prevent wasteful duplication of services in high-density areas while fostering the availability of service in low-density areas. The policy resulted in nearly universal, efficient telephone service at reasonable rates.

Three interrelated factors led to the gradual modification of this policy and the restructuring of the telecommunications marketplace:

1. technological developments
2. customer demand for innovative products
3. FCC and court decisions

Technological Developments

Years of intensive research in electronics—some of which were attributable to the NASA space program—laid the foundation for a variety of practical applications in business and industry. The discovery of the principles of the transistor led to the development of solid-state electronics. Today many industries use solid-state electronics: home appliances, automobiles, and manufacturing, as well as computers and telecommunications.

The many applications of electronics created new products that incorporated both communications and computer capabilities. Telephone switching systems took on many characteristics of computer systems, and telecommunication capabilities became an integral part of distributed data processing systems. The result was the blurring of boundaries between the telephone and computer industries.

Customer Demand for Innovative Products

The applications of solid-state electronics in industries other than telecommunications resulted in a wide variety of popular consumer products. While these applications were highly visible in the marketplace, telephones themselves, private branch exchanges (PBXs), and other telephone equipment did not reflect any apparent changes. Until the mid-1950s all telephones were of the same design, and their color was basic black. Applications of solid-state electronics in the telecommunications industry resulted in faster call completion (direct distance dialing), high-quality transmission, automatic billing procedures, and similar improvements that were invisible and that the public took for granted. These technological improvements generated cost savings for the industry that helped keep telephone rates down. This, too, was accepted as a matter of course.

When designer telephones were finally introduced, they met with enthusiastic acceptance in spite of their higher prices. This suggests that the telephone companies probably should have investigated the potential market before the competitive era forced them to do so.

FCC and Court Decisions

Through the years, the FCC allowed the telephone industry to operate as a total monopoly, thus maintaining the comfortable status quo. The first major challenge to this policy was the Carterfone case. Prior to 1968, telephone customers were prohibited from connecting any device to telephone lines. In the early 1960s the Carter Electronics Corporation began marketing a device called the "Carterfone," an acoustic/inductive device to connect mobile radio systems with the telephone network. Southwestern Bell threatened to discontinue service to customers using the Carterfone, and Carter brought an antitrust suit against the Bell System. The suit was remanded to the FCC, since it had jurisdiction over the conditions under which telephone service was provided. The FCC ruled in favor of Carter, declaring that the Carterfone device filled a communications need. The decision also permitted the interconnection of foreign (nontelephone company) devices to the telephone network. This decision marked the birth of a new industry, which became known as the *interconnect industry.*

The next step toward today's competitive market was the Microwave Communications, Inc. (MCI) decision. In 1963 MCI petitioned the FCC for the right to build a microwave relay system between Chicago and St. Louis. The proposed system would provide common carrier services in direct competition with the Bell System and Western Union. After six years of hearings, MCI was granted permission to build its relay system on the grounds that the data communication needs of the computer industry were not being met by the existing common carriers. The FCC ruled that the microwave system would serve the public interest and that competition would not be harmful. Thus, in 1969 a new area of telecommunications was opened to competition. Since MCI served specialized needs, it became known as a *specialized common carrier (SCC).* In the years that followed, many other companies were authorized to serve as specialized common carriers.

Recent steps toward competition include the Computer Inquiry II decision (1981) and the Modified Final Judgment (1982).

In 1974 the United States Department of Justice commenced litigation against AT&T, charging violation of the federal antitrust laws. Concurrently with the antitrust litigation, the FCC was conducting its own investigation concerning the transmission of data over telecommunication lines. The question under consideration was whether the computer companies were engaging in common-carrier activities and, therefore, should be regulated. The FCC was also considering whether or not AT&T should be permitted to process data during its transmission and thus engage in data processing activities. This investigation led to a decision known as the Computer Inquiry II decision, which specified that:

1. computer companies be permitted to transmit data on an unregulated basis
2. the Bell System be permitted to engage in data processing activities
3. communication services be divided into two categories: basic and enhanced

Basic communication includes dial-up and private-line transmission services. The order specified that these services were to continue to be regulated and provided by the operating telephone companies. It also specified that enhanced services (services that involve computer processing of the transmitted information) and customer-premises equipment be deregulated and provided by a fully separated subsidiary (FSS) of AT&T.

In 1982 the United States Justice Department suit was finally settled out of court by a consent decree that was subsequently modified and approved by the federal court. It is known as the Modified Final Judgment (MFJ). The decree provided that AT&T divest itself of all its operating companies. The operating companies were ordered to transfer all their customer-premises equipment to AT&T; AT&T was to be allowed to provide customer-premises equipment to the public on a deregulated, competitive basis. Later in 1982 AT&T organized American Bell, a nonregulated subsidiary, to market customer-premises equipment in competition with other interconnect vendors. In 1983 American Bell was renamed AT&T Information Systems. The operating telephone companies were directed to confine their activities to the provision of telephone lines and switching services.

The MFJ also stipulated that all long distance service (intrastate as well as interstate) be provided by AT&T rather than the operating companies. The operating companies were also required to provide network access to all long distance companies on an equal basis and are entitled to compensation for that access.

These FCC and court decisions resulted in lowering the barriers to entry in the telecommunications market and encouraging competition. The changes are expected to promote more choices for the American consumer and bring new technology to the market sooner. The decisions indicate that, after years of protecting the communications industry as a total monopoly, the FCC has reversed its former position and embraced competition. A major outcome of the MFJ was the reorganization of the world's largest company, AT&T. *Fortune* magazine described this reorganization as "the biggest American industry has ever witnessed," adding, "A coup this big could fell most governments."

Today's Telecommunication Industry

An important outcome of AT&T divestiture has been a change in the patterns of ownership and servicing of telephone systems. Although users had been authorized to buy and install their own equipment for more than a decade, the majority followed the traditional pattern of obtaining telephones from the telephone company for the payment of a monthly rental fee. In the new era of deregulation, the telephone company is prohibited from providing telephone systems, thus forcing users to select vendors from which to buy telephones.

The new environment also requires customers to be responsible for maintenance and repair of their telephone systems, which were formerly handled by the telephone company.

In summary, both individuals and organizations have been required to assume more responsibility for procuring and managing their telephone systems. Telecommunications management has changed from a passive subscriber role to an active one of sophisticated financial management and decision making. The new environment presents users with more opportunities and more difficult choices.

Knowledgeable experts agree that with informed management, most organizations could reduce their telephone expenses by 20–30 percent without decreasing services. To make the best decisions, managers must understand the new telecommunications environment—the new technology, regulatory climate, various supplier offerings, and cost-saving opportunities available in telecommunications—and be able to match them with their organizations' needs.

Growing Importance of Telecommunications

The 1980s have been described as "the decade of telecommunications," with 1983 marking a turning point in the restructuring of the telecommunications industry. The International Telecommunication Union (ITU), the telecommunications agency of the United Nations, proclaimed 1983 as "World Communications Year" to reflect its growing significance on a global basis.

A study by Frost and Sullivan reported telecommunications operating revenues in the United States as $64 billion in 1980 and predicted that they would reach $119 billion by 1985. Thus, in the five-year period from 1980 to 1985 it is predicted that telecommunications operating revenues will nearly double.

According to the U.S. Bureau of Labor Statistics employment projections for the 1980s, communications will be the fastest growing sector of the economy for the next decade at least.

Careers in Telecommunications

The phenomenal growth that the telecommunications industry is experiencing is creating a need for all types of telecommunications professionals. Never before have telecommunications professionals been in greater demand and shorter supply. Men and women are finding career opportunities with interconnect companies, specialized common carriers, consulting firms, government agencies, the military, service companies (for installation and repair of private telephone systems), telephone answering services, and user organizations as well as with the operating telephone companies. Within these organizations there are opportunities for employment as managers,

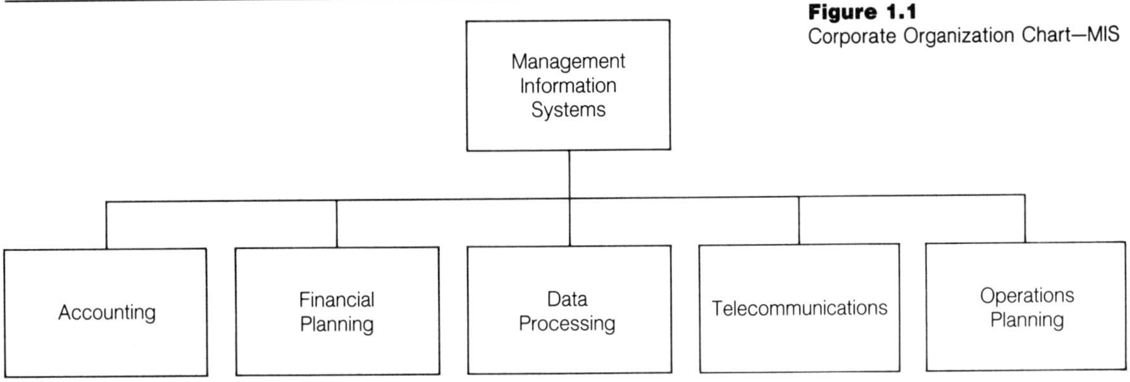

Figure 1.1
Corporate Organization Chart—MIS

analysts, salespersons, engineers, technicians, training specialists, and operators.

Telecommunications plays an important part in the success of any business. All organizations require someone to handle the telecommunications responsibilities. In a small organization, the telecommunications function may be performed by a manager who is also responsible for many other functions, or the responsibility could be delegated to a consultant. Larger organizations generally have a telecommunications manager and a staff of professionals trained in the various aspects of telecommunications. There are many ways in which the telecommunications function could be organized. Figures 1.1 and 1.2 illustrate two possible corporate organization charts. In Figure 1.1 the telecommunications function is in the Management Information Systems group. In Figure 1.2, the telecommunications function is in the Administrative Services group. Chapter 9 of this text examines the responsibilities of the telecommunications manager and the skills required for that position.

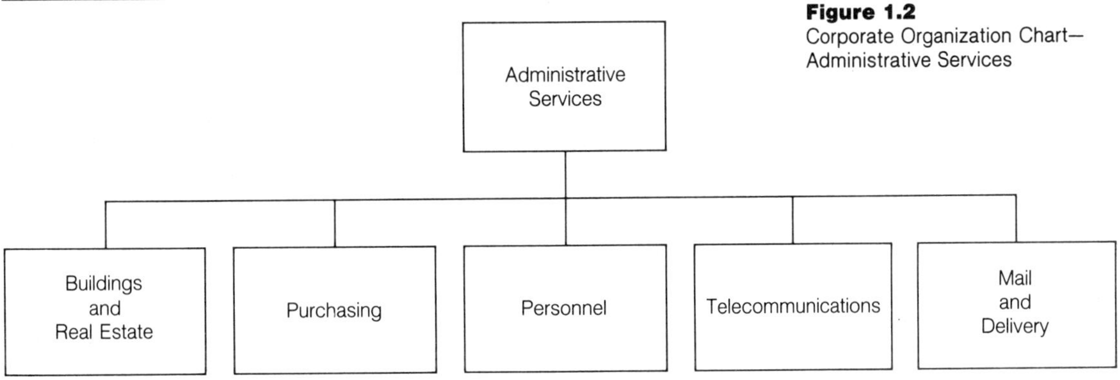

Figure 1.2
Corporate Organization Chart—Administrative Services

Figure 1.3
Career Ladder in Corporate Telecommmunications

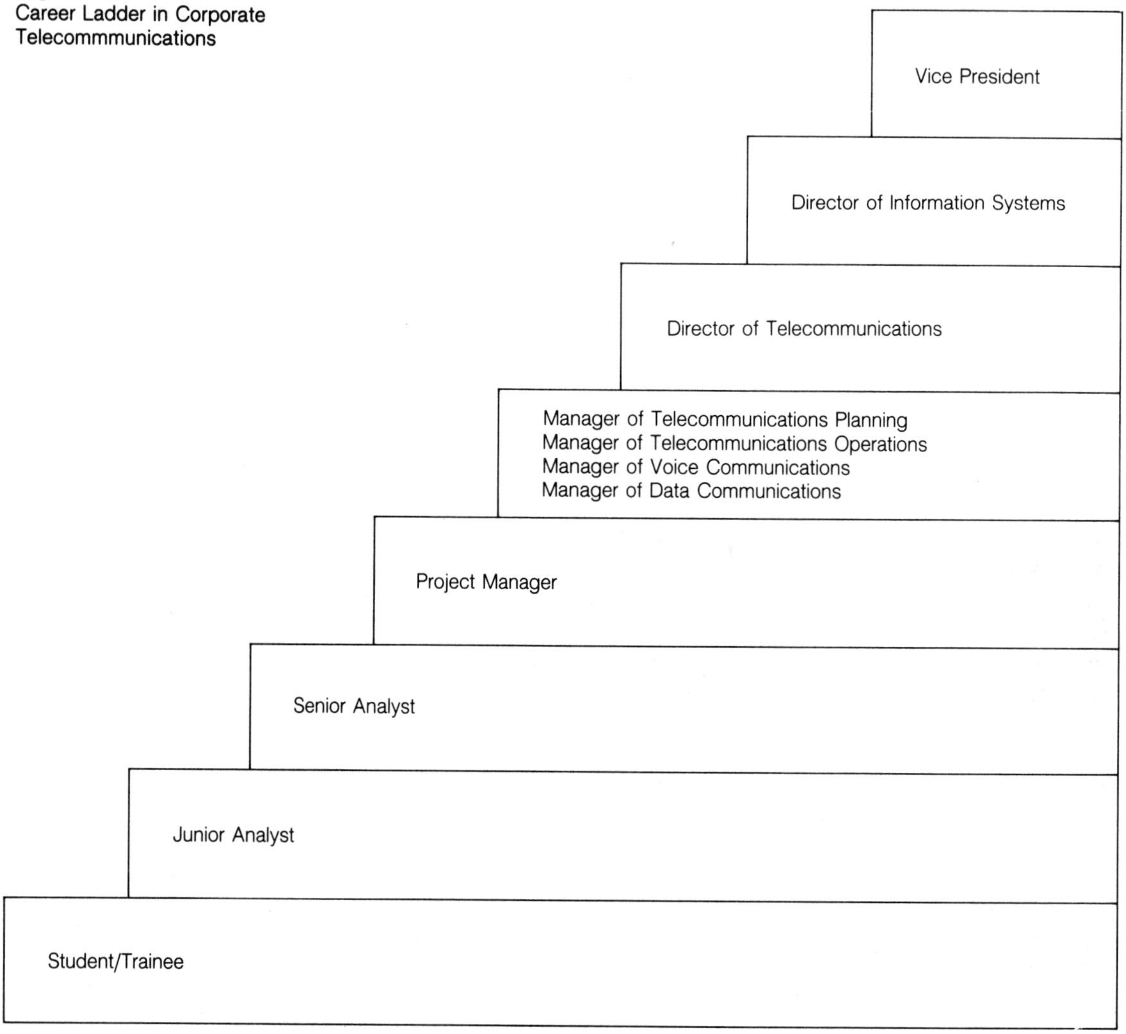

Figure 1.3 illustrates a career ladder that depicts the various opportunities available in telecommunications in a large organization. Similar career progressions exist within interconnect companies and operating telephone companies. Figure 1.4 illustrates the organization of a typical corporate Telecommunications Department.

Education in Telecommunications

For many years, the only way to acquire telecommunications skills was through on-the-job training or industry seminars. Only recently have colleges and universities recognized the need for telecommu-

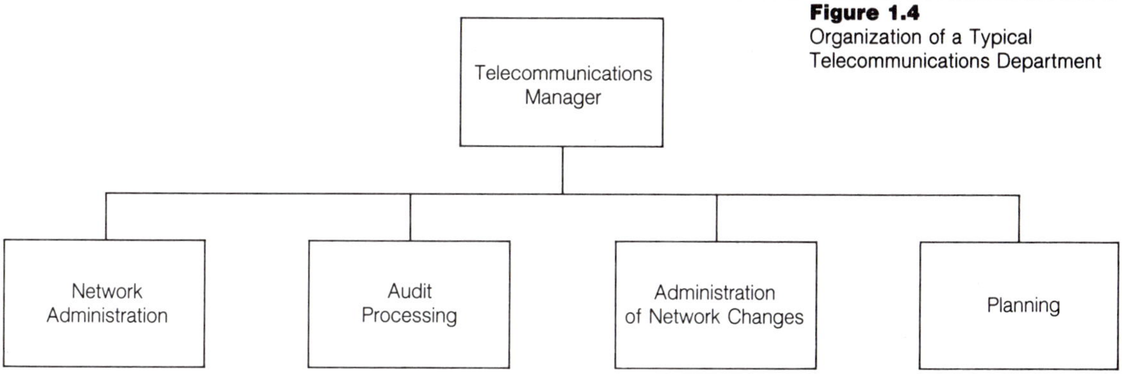

Figure 1.4
Organization of a Typical Telecommunications Department

nications education and begun to offer formal programs in the discipline. The first programs offered were in engineering curricula, based in electrical engineering with a telecommunications specialty. More recently, university programs have begun to address the needs of telecommunications management.

As a general rule, telecommunications programs lead to a baccalaureate or graduate degree. However, to meet the needs of professionals who are not interested in degree programs, some colleges and universities now offer certificates for series of telecommunications courses.

An analysis of an International Communications Association (ICA) survey of the status of graduates of telecommunications curricula in industry concluded that "Telecommunications programs, as represented in the 1980 survey, appear to be well-received in industry, engaging in technology-specific managing and engineering functions, and drawing the attendant high salaries."[2]

In addition to complete programs in telecommunications, many community colleges, colleges, and universities have implemented introductory courses in telecommunications in their information management, computer science, electronics, communications, office management, secretarial science, and other curricula.

Summary

This chapter introduced the reader to the field of telecommunications and its role in moving information. The marriage of computers and communications technology has enlarged both fields, making

2. Jane M. Clemmenson, "Telecommunications Curricula: the Status of Program Graduates in Industry," *Telecommunications*, September 1982, p. 101.

possible distributed processing, immediate accessing of centralized data bases, and a wide array of enhanced telephone service features.

Technological innovations, industry deregulation, and the new competitive environment have created an urgent need for telecommunications management. Phenomenal industry growth has resulted in the need for all types of telecommunications professionals.

In succeeding chapters, telecommunications will be discussed in relation to, respectively:

1. early history
2. industry structure and regulation
3. telephony
4. networking
5. data processing
6. data communications
7. services available
8. managing a system
9. selecting and implementing a new system
10. traffic engineering
11. rate making
12. the new competitive environment

Information on professional organizations, industry seminars, and equipment suppliers as well as a glossary and index can be found in the back of the book.

Review Questions

1. What are the two characteristics that identify telecommunication systems?
2. Name four types of telecommunication systems available today.
3. Name several industries whose very existence depends upon telecommunications and explain why this is so.
4. How does telecommunications make distributed processing possible?
5. What are some of the types of organizations in which telecommunications career opportunities can be found?
6. What are some types of positions available in telecommunications?

References and Bibliography

Bly, Robert W., and Gary Blake. *Dream Jobs: A Guide to Tomorrow's Top Careers*. New York: Wiley & Sons, Incorporated, 1983, 154–75.

Clemmensen, Jane M. "Telecommunications Curricula: The Status of Program Graduates in Industry." *Telecommunications,* September 1982, 99–101.

Harlow, Ronney. "Growing Choices Creating Urgent Need for More Telecom Professionals." *Communications News,* March 1983, 30–31.

International Communications Association. "Major ICA Study Finds 'Typical' Manager." *Communications News,* January 1983, 48–51.

Kander, Sharon L. "Women in Telecommunications." *Teleconnect,* May 1983, 30.

LaBlanc, Robert E. with Richard M. Wolf and Elizabeth A. LaBlanc. "Communications + Data = Compunications." *Telephone Engineer and Management,* May 1, 1983, 67–71.

Paul, Carolyn. "College and University Courses Begin to Reflect Needs of Telecommunications Management." *Communications News,* September 1980, 28–29.

Powers, K. H. "Communication—A Mild Explosion." *RCA Engineer,* Nov./Dec. 1980, 4–7.

Roberts, Bert C., Jr. "The Information Economy." Address before the National Conference on Telecommunications Education, Dearborn, Michigan, January 17, 1984.

Trainor, Timothy N. *Computer Literacy Concepts and Applications.* Santa Cruz, Calif.: Mitchell Publishing, Inc., 1984.

Wiley, Don. "World Communications Year—1983—Will Be Landmark Year in USA." *Communications News,* January 1983, 26–41.

———. "Largest Restructuring Ushers in New Year and New Area of Rapid Telecom Growth with Vast New Opportunities." *Communications News,* January 1984, 28–39.

Williams, Frederick. *The Communications Revolution.* Beverly Hills, Calif.: Sage Publications, 1982.

2 Early History of Telecommunications

The principal form of communication between people throughout the ages has been speech, a process which, until the nineteenth century, was limited by the distance between people. Early civilizations used coded light signals to transmit information over considerable distances, but it was the transmission of speech over distances that stirred people's imaginations. It wasn't until the nineteenth century, however, that the electromagnetic mode of speech transmission came into use.

Given the historical nature of the telecommunication industry's structure and operating methods, a perspective of its development will be useful in clarifying the dramatic changes taking place in the industry today. This chapter highlights the early history of telegraphic communications, the formation of the Western Union Telegraph Company, the impact of telegraphic communications upon the political and business organizations in the United States, the early history of telephony, the formation of the Bell System, and the interrelationship between these two industries from the 1880s through the 1920s.

Early Methods of Communication

Humans are social beings; one of our basic drives is to communicate with others. Familiar expressions such as "get it out in the open," "get it off my chest," and "talk it over" reflect our need for communication as a means of making us feel better. Psychiatrists, ministers, and counselors promote mental health as they help us "talk things over."

Businesses use a variety of methods to encourage two-way communication. Periodic consultations between employees and their supervisors; company-sponsored social events such as picnics, bowling teams, and Christmas parties; and business conferences, all serve to promote upward and downward communication among employees at various organizational levels.

Nonverbal Communications

People have always conveyed information to each other in a variety of ways. In earliest times communication consisted primarily of signs, gestures, and facial expressions. These elementary, nonverbal methods, used originally to beckon, to warn, to approve, or to disapprove, are still effective means of communication. The friend waving to us from across the room, the police officer on the corner signaling traffic, the baseball umpire behind the plate, and the pedestrian hailing a taxicab, all use gestures to relay information to others.

Current interest in body language—the way we express ourselves through our postures and movements—has called attention to nonverbal communications. Some authorities on the subject claim that at least 50 percent of the impressions people convey result from nonverbal exchanges. Although some body actions, such as crossing the arms or legs, are subject to different interpretations, certainly most of us would regard smiling as friendliness, nodding the head as agreement, raising the eyebrows or wrinkling the nose as skepticism, tapping the feet or drumming the fingers as impatience, and frowning as discontent.

Our first use of sound to communicate did not involve the human voice, but a variety of beating or tapping noises to attract attention. These sounds were generally produced by hitting the ground or another surface such as a hollow log. By striking with their fists or with a small mallet, early people could produce sound waves that traveled in all directions from the source. Later, our ancestors progressed to vocal but nonverbal means of communication such as cries, screams, groans, and squeals. These primitive methods, universally recognized as spontaneous expressions of emotion, continue to be used to convey the same meanings today.

The next step—the development of a system of speech—was one of the most important milestones in the history of human communication. Spoken language, with its many words, intonations, and inflections, introduced into the exchange of information a degree of precision previously unknown. Ever since humans first learned to talk, spoken language has been our principal means of communication. Its chief limitation was that the talker and listener had to be "within earshot."

Through the years people developed a number of long-range communication methods, including megaphones, smoke signals, oil lamps, bells, whistles, and semaphores. Many of them are still used today. For example, bells mark the time of day, invite us to worship, and announce football victories. The semaphore, an apparatus for signaling by means of flags, lights, or mechanical arms, is still used to supplement electronic methods in ship, railway, and air communications.

It was the transmission of spoken language over distances, however, that continued to excite people's imagination over the centuries. Finally, with the discovery of electricity in the early nineteenth century, the essential elements for long distance sound transmission were available.

Early History of Telegraphic Communications

The introduction of the electric telegraph by Samuel Findley Breeze Morse revolutionized communications. Morse, who earned his livelihood as a portrait painter and professor of the Literature of Arts and Design at New York University, had always been interested in science and invention. His interest in electricity was heightened when, during a return voyage from Europe, he watched a fellow passenger demonstrate how a piece of iron became magnetized by electric current. This demonstration led to discussions among other passengers of Michael Faraday's recent publication on magnetism.

The Morse Telegraph

Upon returning home, Morse constructed telegraph sending and receiving instruments and formulated the principles of his dot-dash-space-code based on the duration or absence of electrical impulses. He developed his code by assigning dots and dashes to the letters of the alphabet, with the most frequently used letter having the simplest code. Thus, Morse code represents the frequently used letter *e* by a single dot, which requires the least electricity and transmission time; the infrequently used letter *x* is represented by dot-dash-dot-dot, which requires the most electricity and transmission time. Morse code consists solely of combinations of dots and dashes. A very short burst of electric current represents the dot and a slightly longer burst represents the dash.

Prior to the development of telegraphy, the only long distance communication was by postal service. Early mail service delivered letters by horseback, stagecoach, and steamboat. Delivery schedules were infrequent and often unreliable. The inadequacy of long distance methods of communication in those days was poignantly brought home to Morse in 1825. He had been commissioned by the

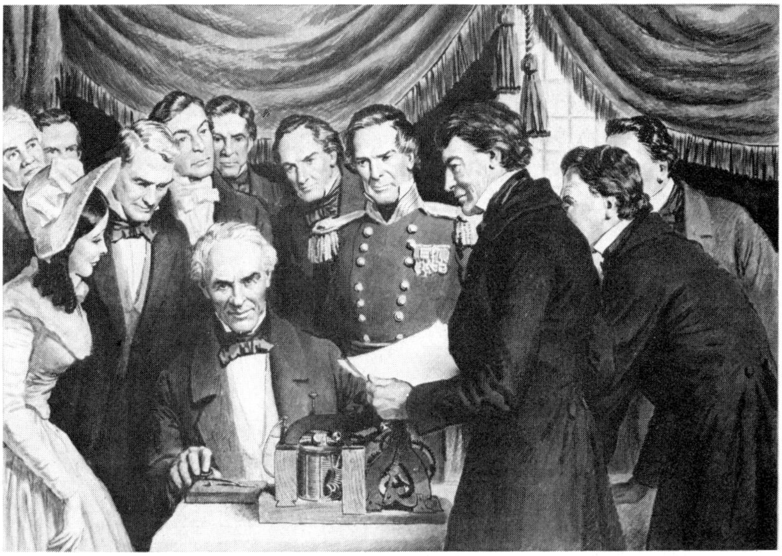

(Courtesy Western Union)

Figure 2.1
Morse Sends the First Telegram

City of New York to paint a portrait of the Marquis de Lafayette, an appointment that took him to Washington. While there, he received word of the sudden death of his young wife back home in New Haven, Connecticut—word which, because of the slowness of communications, did not reach him until twenty-four hours after her funeral!

On September 2, 1837, Morse demonstrated his telegraph to a group of professors and friends. This exhibition suggested the practicability of the invention and resulted in the enlistment of two partners, chemistry Professor Leonard Gale and a young inventor, Alfred Vail, to help promote the telegraph. Late in September 1837, Morse filed a caveat for his invention in the United States Post Office.

In February 1838, Morse demonstrated his telegraph before President Martin Van Buren and his cabinet, hoping to have the telegraph accepted for government use. However, it wasn't until March 3, 1843, that Congress appropriated funds to test its workability. An experimental telegraph line was constructed between Washington and Baltimore. On May 24, 1844 (see Figure 2.1), Morse sent the first public telegram over this 40-mile line, transmitting the message, "WHAT HATH GOD WROUGHT!"

In spite of his successful demonstration, the United States postmaster general decided that the telegraph was a toy that probably would not be successful commercially, and the government withdrew its support.

Commercialization of the Telegraph

Morse then enlisted private capital and in May 1845 organized the Magnetic Telegraph Company, the first telegraph company in the United States. By 1851, 50 telegraph companies using Morse telegraph patents were in operation in the United States, each serving a different section of the country. Accordingly, when a user wished to send a telegram over a long distance, it had to be retransmitted from company to company. This procedure not only slowed down delivery but also made it difficult to pinpoint responsibility when poor service resulted.

The various telegraph companies were highly competitive; some prospered while others had difficulty staying in business. In time it became apparent that consolidation of the companies would promote better service. However, petty jealousies blocked attempts at organizing the principal telegraph lines under the Morse patents into an association.

The Telegraph's Impact on Politics and Business

The telegraph soon exerted a strong influence upon the political and economic life of the nation. Its impact upon railway operation and management was substantial. The telegraph provided the railroads with electric train dispatching, informing management of the location of every train on its rail system. The railroads, for their part, provided the telegraph companies with an exclusive right-of-way; telegraph wires were strung on poles beside the railroads. Since each industry had something of value to offer the other, their service contracts with each other became valuable assets.

Another industry closely associated with the development of the telegraph was the newspaper industry. The advent of the telegraph revolutionized the collection and dissemination of news. Since the industry was dependent upon receiving and printing the news as rapidly as possible, it had to employ the only method of rapid communication available at the time—the telegraph. The high cost of individual telegraphic services suggested the need for a cooperative news reporting service. As a result, the New York papers organized the New York Associated Press.

In return for dealing exclusively with certain leading telegraph companies and for promising not to print anything detrimental to these companies, the Associated Press received reduced rates on its business. Telegraph companies, too, benefited from this agreement.

Both state and federal government officials depended upon the telegraph in carrying out their responsibilities. Politicians and government officials used the telegraph to keep in touch with their constituents, and the people at home used it to inform their representatives of their wishes with respect to legislative policies.

(Courtesy Western Union)

Figure 2.2
Early Telegraph Office

The Civil War Years

By the beginning of the Civil War, telegraph lines linked most of the major cities in the nation (see Figure 2.2). The war years saw a dramatic rise in the use of telegraphic communications. The military, faced with assembling troops and supplies and moving them to the battle fronts, found the telegraph invaluable. Businesses, too, deluged the telegraph offices with messages in a desperate effort to get their affairs in order before contacts between the North and the South were broken.

The heavier concentration of railroads and telegraph lines in the northern states gave the Union forces a substantial advantage over the South during the Civil War. Further, the telegraph companies located in the North were relatively secure from the ravages of war, while those whose lines extended from north to south found many of their lines destroyed. Thus, some telegraph companies emerged from the war years in improved economic condition while others were barely able—or unable—to remain in business.

One of the telegraph companies whose business prospered during the war years was Western Union. Originally incorporated under the name New York and Mississippi Valley Printing Telegraph Company, the organization was reincorporated in 1856 as the Western Union Telegraph Company. With its telegraph lines located north of the Mason-Dixon line and the Ohio River, it was situated

Figure 2.3
Alexander Graham Bell in 1876, the Year the Telephone Was Invented

(Reproduced with permission of AT&T)

advantageously to profit from the increase in telegraph business brought about by the war.

When General George B. McClellan in 1861 took over the Department of the Ohio, which included western Virginia, Ohio, Indiana, Illinois, and later, Missouri, he appointed Anson Stager, general superintendent of Western Union, to administer for military purposes all telegraph lines within his department. Stager proceeded to coordinate the operation of the military and commercial telegraph lines within his jurisdiction into a highly effective system.

Western Union took advantage of its good fortune to improve its position within the industry. Gradually, it absorbed all of the competing independent telegraph companies. With the acquisition of the United States and American Telegraph Companies in 1866, Western Union emerged as the nation's largest corporation and its first powerful monopoly.

Early History of Telephony

On March 7, 1876, U.S. Patent No. 174,465 was issued to Alexander Graham Bell (Figure 2.3) for his invention of the telephone. This patent, often described as "the most valuable single patent ever issued," laid the foundation for an industry that serves nearly every neighborhood and home in the land.

"Mr. Watson, Come Here, I Want You!"

These historic words, shouted by Alexander Graham Bell to his assistant, culminated years of experimentation in sending electric messages over wires. "I can hear, I can hear the words," Watson exclaimed excitedly as he rushed into Bell's workshop. The telephone had talked!

Like his father and grandfather, Mr. Bell was a teacher of elocution, or speech transmission. His research in the physiology of speech was no doubt influenced by both his father, the inventor of "Visible Speech," a written code used in training deaf persons to speak vocally, and by his mother, an accomplished musician whose own loss of hearing intensified his interest in the electric transmission of vocal messages. Bell earned his livelihood teaching classes for the deaf; at night he worked on his experiments. Bell postulated that if sound could be converted into electrical signals, it should be possible to transmit speech over a distance electrically. By employing a diaphragm to produce an undulating current, he proved his theory to be right. Thus, he learned the fundamental principle that enabled him to develop the telephone.

At the same time that Bell was conducting his experiments, Elisha Gray, an expert electrician and well-known inventor in the field of telegraphy, was independently working on the same problem. Both Bell and Gray attempted to construct a device to transmit a number of telegraph messages over a single wire simultaneously using interrupted tones of different frequencies, a concept known as *harmonic telegraphy.* Bell approached the problem through his knowledge of acoustics; Gray's approach was through electricity. Each concluded that a combination of harmonics could be sent over a wire simultaneously; this discovery led each of them to postulate further that the human voice could generate impulses that could be transmitted over wires.

Bell's application for a U.S. patent was filed in Washington by his attorney on the morning of February 14, 1876. (See Figure 2.4.) For the lack of a better name, he called his invention "an improvement in telegraphy." A few hours later on the same day, Elisha Gray came to the same patent office and filed a caveat—a notice of intent to perfect his ideas to file a patent application within three months—also for an electric telephone. Since Bell's papers were the first to be filed, a patent for the telephone was issued to him on March 7, 1876. However, at this time no one had transmitted a single intelligible sentence by the telephone.

Figure 2.4
1876 Liquid Telephone

(Reproduced with permission of AT&T)

Bell and his assistant, Thomas A. Watson, pressed on with their experiments to test the workability of the theory described in the patent application. At Bell's direction, Watson built a variable resistance transmitter that used sulphuric acid as a conductor of electrical current. The transmitter was set up in Bell's workshop (see Figure 2.5) and connected by a wire to a receiver in his bedroom. When the device was ready for testing, Bell adjusted the transmitter while Watson went into the bedroom and put the receiver to his ear. Almost at once he heard Bell's voice saying excitedly, "Mr. Watson, come here, I want you!" Watson rushed down the hall and found that Bell had upset the sulphuric acid, spilling it all over his clothes. Thus, on March 10, 1876, three days after the patent had been issued, Bell had developed a working telephone.

Financial backing for Bell's experiments was provided by the father of two of his students, Thomas Sanders, a successful leather merchant, and Gardiner Greene Hubbard, a prominent attorney. Sanders and Hubbard had furnished the money for Bell's experiments in return for an equal share in any patents obtained. However, they

(Reproduced with permission of AT&T)

Figure 2.5
Alexander Graham Bell's Original Laboratory

were anticipating his developing a harmonic telegraph, not a telephone. They had little interest in the telephone invention and were skeptical about its possibilities for commercial use. In the fall of 1876, they offered to sell all of Bell's patents to Western Union Telegraph Company for $100,000, but the offer was refused.

Meanwhile, Bell continued his experiments to improve telephone performance, hoping to ultimately realize a profit from his invention. On January 15, 1877, he filed an application for a patent for the *box telephone,* an improved version of his original telephone instrument (see Figure 2.6). The patent was issued on January 30, 1877. Bell's assistant, Thomas A. Watson, also continued to experiment and developed further telephone improvements including ringers (bells) and switchboards.

Publicizing the Invention

In the summer of 1876 Bell exhibited his "speaking telephone" at the Centennial Exposition in Philadelphia. It attracted little attention until Bell demonstrated it to Dom Pedro, Emperor of Brazil (Figure 2.7), and Sir William Thompson, British physicist (later Lord Kelvin). "My God, it speaks!" said Dom Pedro, and Sir William Thompson,

Figure 2.6
The Box Telephone

(Reproduced with permission of AT&T)

Figure 2.7
Bell Demonstrates His Invention to Dom Pedro

(Reproduced with permission of AT&T)

after careful scientific scrutiny, said it was the greatest thing he had seen in America.

On February 12, 1877, Bell gave a lecture before a well-known scientific society, the Essex Institute, at Salem, Massachusetts. The society members there were especially interested in the telephone because Bell had performed his early speech and sound experiments in Salem. This first lecture was free to members of the society, and it created so much interest that Bell was asked to repeat it—this time for an admission fee. Again, the house was filled and the lecture was well received. At the first public demonstration, a newspaper correspondent from the *Boston Globe* telephoned a report of the lecture to his editor. This was the first use of the telephone in news reporting.

Spurred on by the public interest in his lectures and demonstrations, Bell assembled a presentation resembling a vaudeville show. He rented public halls and, assisted by Watson, entertained audiences by transmitting conversations and songs over the telephone. To add interest, he permitted members of the audience to talk to someone over the telephone. The demand for the lectures was fortuitous for Bell because it solved a temporary money problem. Bell had several urgent reasons for needing money. In addition to establishing a telephone company, he had fallen in love with and wished to marry Mabel Hubbard, Gardiner Hubbard's daughter who had been deaf since early childhood. During the next five months, Bell and Watson delivered lectures in New York and a number of New England cities. On July 11, 1877, Bell and Mabel Hubbard were married. Shortly thereafter, the couple sailed for England, taking with them a complete set of telephones.

The Trusteeship

In July 1877 the telephone industry was formally organized by Hubbard with the creation of the Bell Telephone Company. The first organization was a trusteeship composed of Alexander Graham Bell, Gardiner Hubbard, Thomas Sanders, and Thomas A. Watson. With Hubbard acting as trustee, the organization began to manufacture and install telephones.

Hubbard, an attorney, had observed that one of his other clients, the Gordon-McKay Shoe Machinery Company, had built a highly successful business organization with its policy of leasing equipment instead of selling it. The company leased the shoe-sewing machines to shoemakers, retaining their title to the machines and receiving a royalty for every pair of shoes sewed with the machines.

By the powers vested in him, Hubbard made a decision that was to have far-reaching effects—the decision to rent telephones instead of selling them. This resulted in the sale of service only, which became the basic principle of the telephone industry. In spite of the critical need for business capital, Hubbard held firmly to the leasing principle as the basis for the development of the telephone business. This policy was a major factor in the financial success of the Bell system; later it became the policy of other telephone companies as well.

The trustees also authorized a system of franchises whereby agents in various parts of the country could provide telephone service by paying a license fee to the trusteeship. In time, the trusteeship was supplanted by a corporate organization.

Early Telephone Service

The first telephone instrument bore little resemblance to the telephone that we know today. It was a crude, cumbersome apparatus consisting of a device that served both as a transmitter and receiver, and a connecting length of wire. In order to use the instrument, the user had to shift it back and forth between the mouth and ear. Occasionally, an affluent customer would obtain two instruments in order to use them simultaneously. One of the first telephone's biggest drawbacks was that it had no bell or call signal; thus, there was no way for a person to know that a call was waiting.

By the fall of 1877 there were about six hundred telephone subscribers, all using private lines. Each telephone was connected to another telephone by a direct line consisting of a single iron wire. There were no central exchanges and no switchboards; conversations could take place only between two telephones at each end of the line. On January 28, 1878, the first telephone exchange was opened in New Haven, Connecticut, serving 21 subscribers. The early exchanges served only a few customers, and calls were completed manually by an operator sitting at a switchboard. Calls were "put through" by name rather than by number. The exchange made it possible to connect any telephone with any of the other telephones in the exchange, greatly increasing the usefulness of the telephone. Increased demand for telephone service soon resulted, and within a few months a number of telephone exchanges were opened throughout the country.

On February 12, 1878, the New England Telephone company was formed. This was a licensing, not an operating, company. It held an assignment of rights to the Bell patents for New England and authorized agents to provide telephone service in return for the payment of a license fee.

On March 20, 1879, both the New England Telephone Company and the Bell Telephone Company were consolidated under the name National Bell Telephone Company.

Emergence of Competition

Meanwhile, Western Union, which had been a successful telegraph company since 1856, reconsidered the prospects of telephony. Already national in scope, Western Union had an extensive network of wires connecting its offices in hotels, railway stations, and other public places that could be used as a nucleus to provide telephone service.

After Bell's patents for his telephone invention had been issued, patent applications for various forms of speaking telephones and transmitters were filed with the United States Patent Office by different individuals. Prominent among these were Elisha Gray of Chicago; Thomas A. Edison of Menlo Park, New Jersey; and Professor Amos E. Dolbear of Somerville, Massachusetts. Western Union purchased the Gray, Edison, and Dolbear patents and organized its own telephone company, the American Speaking Telephone Company. Ignoring the Bell patents, it began offering telephone service to the public. A period of intense competition followed. In many instances, both the Bell Company and the American Speaking Telephone Company established exchanges in the same town, which resulted in two telephone systems that were not interconnected.

Vail's Leadership

To counter this attack, the Bell Company leaders did two things. First, they hired a professional manager, Theodore N. Vail (Figure 2.8), to manage their organization. Second, they filed a lawsuit against Western Union for infringement of Bell's patents.

Vail had been superintendent of the Post Office's Railway Mail Service, where he was recognized as an outstanding business manager. He left this secure, well-paying job to become the Bell Telephone's first general manager in July 1878 and to face the challenge of directing the struggling young company. When Vail took over, there were only 10,755 telephones in service, competition was intense, and the Bell Company was faced with many technical problems. Vail brought to Bell the management expertise the company so badly needed. His leadership contributed much to the success of the Bell Company and led to the eventual formation of the Bell System.

Patent Litigation

In September 1878 the Bell Company filed a suit in the Circuit Court of the United States, District of Massachusetts, against the giant Western Union Telegraph Company—technically against Peter A. Dowd, agent for the Western Union's telephone subsidiary—for

Figure 2.8
Theodore N. Vail (about 1885)

(Reproduced with permission of AT&T)

infringement of the Bell patents. Western Union engaged George Gifford, a prominent patent attorney, as its chief counsel in the case.

After investigation, Gifford became convinced that the Bell Company would win and advised settlement of the suit. In November 1879 the two parties reached an out-of-court settlement. The settlement provided that Western Union withdraw from telephone service and sell its network and patents to the Bell Company. In return, Bell agreed to stay out of the telegraph business and to pay Western Union 20 percent of its telephone rental receipts over the 17-year life of the patents. This agreement added 56,000 telephones in 55 cities to the Bell Company. Over the life of the agreement, Bell paid Western Union approximately $7 million.

In March 1880, the American Bell Telephone Company, successor to the National Bell Telephone Company, was formed to carry on the consolidation of Bell and Western Union properties. This company remained parent company of the Bell System until December 30, 1899.

Formation of American Telephone and Telegraph Company

In 1885 a new company, the American Telephone and Telegraph Company (AT&T), was formed to build and operate long lines and render nationwide telephone service. The long distance lines interconnected the regional companies that had developed through the franchise agreements. For the first 15 years of its existence AT&T was a subsidiary of American Bell Telephone Company and was generally called the Long Distance Company. Theodore N. Vail became the first president of the American Telephone and Telegraph Company, a post he held until his resignation in 1887.

In 1900 AT&T absorbed the American Bell Company and became the headquarters company of the Bell System. It continued to provide long distance service through its Long Lines Department.

In 1911 AT&T consolidated the operations of the franchise companies into state or territorial units. These territorial units became the structure known as the Bell Associated Companies. Each of the companies paid a license contract fee to AT&T to cover costs of development of new equipment and improved telephone service. The license fee replaced the royalty fee that franchise companies had paid for the use of the Bell patents.

Early Telephone Manufacturers

The first telephones were manufactured in the electrical shop of Charles Williams, Jr., in Boston, where Bell and Watson had conducted many of their early experiments. As the demand for telephones increased, other small shops were licensed to manufacture telephones and related equipment to Watson's specifications. In the next few years telephones and telephone equipment were made by six manufacturers, each producing equipment of differing design and quality. It soon became apparent that a centralized source of high-quality, standardized equipment was needed.

The largest electrical manufacturer in the United States was the Western Electric Manufacturing Company of Chicago, the company that had supplied Western Union's telephone equipment. It had been organized in 1872 as successor to Gray and Barton, manufacturers of electrical equipment, including telegraph apparatus and fire and burglar alarms. From the first, the company gained a reputation for quality workmanship. In 1881 the Western Electric Manufacturing Company was reorganized to enfold some other telephone instrument and switchboard manufacturers.

Since Western Electric had pioneered in electrical equipment and telephone apparatus, the company was well qualified to manufacture Bell telephone equipment. On February 6, 1882, the Bell Company purchased the Western Electric Company and it became Bell's manufacturing unit and the sole supplier of Bell telephone equipment. Ownership of Western Electric gave the Bell System assurance of standardized equipment of high quality, economies of scale, and a dependable supplier.

The Independent Telephone Companies

The early years of the telephone's life saw dramatic improvement in service, rapid growth in the demand for telephones, and the advent of competition. As the value of the telephone became increasingly apparent, many persons appeared who claimed that they had developed instruments that would transmit speech over a distance electrically. Numerous legal battles over patent rights resulted. During the life of the Bell patents, the Bell Company was involved in over 600 lawsuits; it won every one.

The growing interest in telephones caused many people to regard the telephone industry as an attractive business opportunity. As a result, many new telephone companies were formed. Before the expiration of the Bell patents, more than 125 competing companies were in operation. Some of the new companies were organized under a Bell franchise agreement; others began operation "independent" of any Bell affiliation and in direct competition with the franchised companies. The term *independent telephone company* is still used today. Bell's defense against these competing companies took two directions:

1. litigation involving patent infringement
2. refusal to interconnect independents' subscribers with Bell facilities

After the expiration of the Bell patents in 1893 and 1894, more independent telephone companies entered the market. By the early 1900s independent telephony became a serious threat to the Bell System. The Bell System generally concentrated on serving the larger cities where potential customers were plentiful; the independents concentrated on serving small communities and rural areas ignored by Bell. However, sometimes two companies established facilities within the same locality. Since the two systems were not interconnected, a customer was limited to calling only those subscribers served by the same telephone company. In order to have contact with a customer of the other telephone company, it was necessary to subscribe to the other service. Businesses, particularly, found this impractical since they were unable to serve an entire community unless they had two telephones and two directories.

Formation of a National Trade Association In June 1897 delegates from independent telephone companies met in Detroit, Michigan, to establish a national trade association for the industry. They adopted the name Independent Telephone Association of America and dedicated their efforts toward addressing mutual problems and strengthening their segment of the industry.

From 1897 to 1915 the association underwent many name changes and mergers, including one with the independent manufacturers' association. Finally, in 1915, the United States Independent Telephone Association (USITA) was chartered, incorporating the previous organizations. During this period of growth and competition, USITA played a vital role in uniting the industry and providing information and advice on many subjects relating to telephony. For many years it has been recognized as the national voice of the independent telephone industry.

In October 1983 USITA dropped the word "independent" from its name and became U.S. Telephone Association (USTA) to attract Bell operating companies (BOCs) to its membership.

Independent Manufacturers Prior to the expiration of the Bell patents, many independent experimenters were working on improvements to the telephone. One of the most spectacular developments was the dial system constructed by Almon B. Strowger, a Kansas City undertaker. His original instrument (Figure 2.9) had five pushbuttons lined up in a row. To call number 21, for example, the user simply pushed the first button twice and the second button once. Strowger later introduced a large dial in place of the buttons (Figures 2.10 and 2.11). Strowger's first unit went into service in La Porte, Indiana, and its use spread rapidly among independent telephone companies. Later, the Bell System also adopted it.

The independent manufacturing and operating companies also pioneered a number of other developments, including a handset telephone, the forerunner of the instrument in use today. It contained a transmitter and receiver in the handle and was connected to the telephone by a cord.

The larger independent telephone companies own their manufacturing and supply companies. The largest independent telephone company, General Telephone and Electronics (GTE), owns two manufacturing subsidiaries: Automatic Electric and Lenkurt Electric. These subsidiaries manufacture communications equipment for sale not only to GTE companies but also to other firms and governments in the United States and abroad.

Vail Returns

Competition had caused a severe financial drain on AT&T; between 1902 and 1907 its debt grew from $60 million to over $200 million. In 1907 Theodore N. Vail came out of retirement, at the age of 62, to assume the presidency of AT&T for the second time during a period in which the company was in critical condition. In addition to being on the verge of bankruptcy, its service had deteriorated, and its public image was very poor.

In an effort to make AT&T the sole supplier of telecommunications services in the United States, Vail continued the policy of buying

Figure 2.9
Strowger Switch

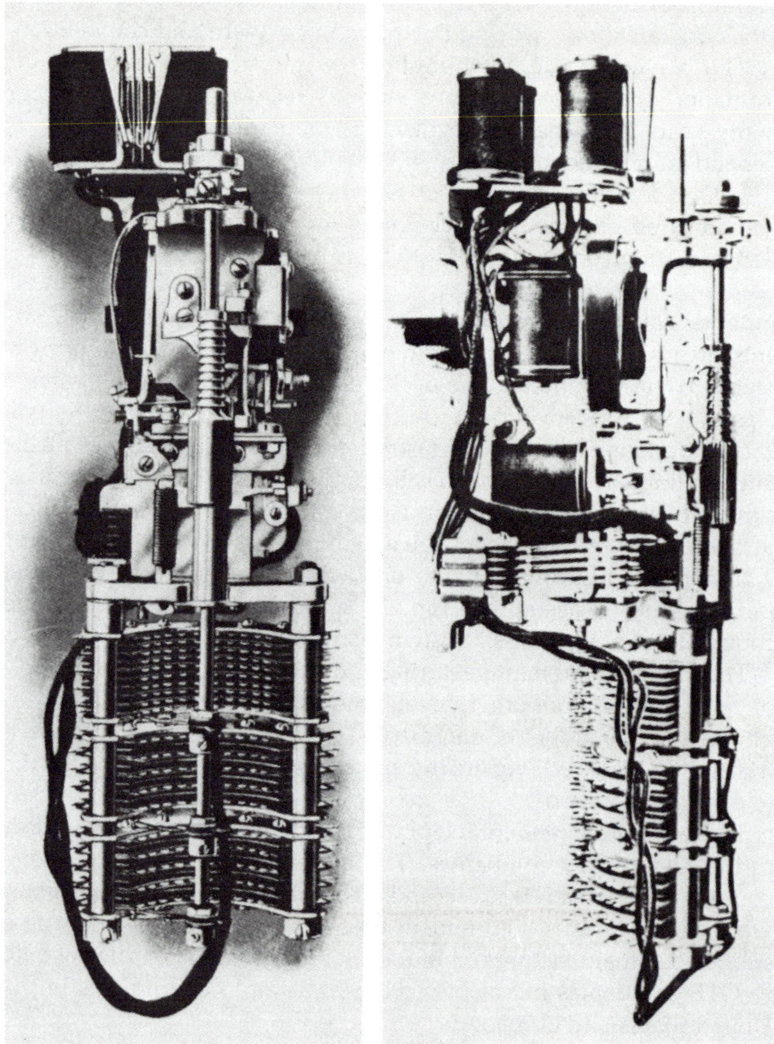

(Courtesy of Bell Laboratories)

independent telephone companies. He also decided that it would be advantageous for AT&T to get into the telegraph business. Accordingly, in 1909 AT&T bought 300,000 shares of Western Union stock, enough to give it working control. In 1910 Vail became president of Western Union, making him president of both companies.

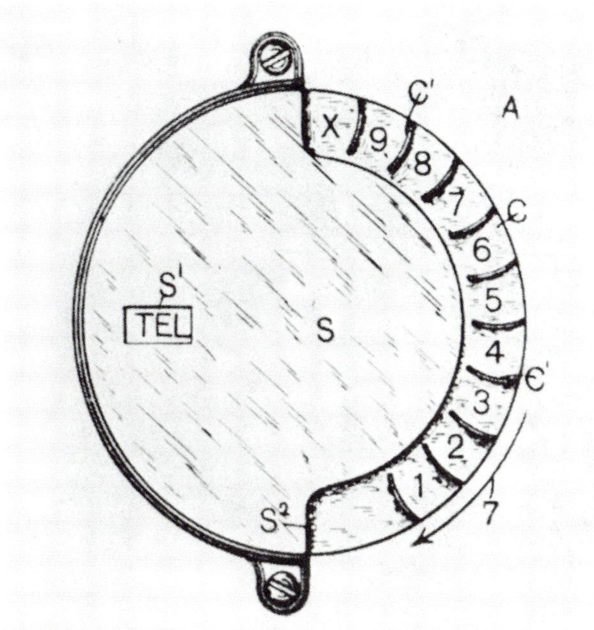

Figure 2.10
Strowger Finger Wheel Dial

(Courtesy of Bell Laboratories)

The Kingsbury Commitment

During the first decade of the twentieth century, AT&T had employed these tactics to combat the independents:

1. rate cutting
2. buying out competitors
3. refusing to interconnect

The most powerful weapon was the refusal to interconnect. Since Bell, under patent protection, had developed an extensive long distance network, its service was of more value than that of the independents who had few long distance facilities.

In time, the ruthless competition between telephone companies and the drawbacks of duplication of telephone facilities became apparent to federal and state legislators. In 1910 Congress took the first step toward industry regulation, placing certain aspects of interstate telephone operations under the jurisdiction of the Interstate Commerce Commission. The commission was given jurisdiction over telephone companies for just and reasonable charges, passes and franks, preferences and prejudices, filing contracts, reports to the commission, investigations, furnishing information, joint rates, uniform system of accounts, and preservation of records.

In 1912 the independent companies protested to the Department of Justice that the Bell organization, now controlled by AT&T, was

Figure 2.11
Strowger Wall Set

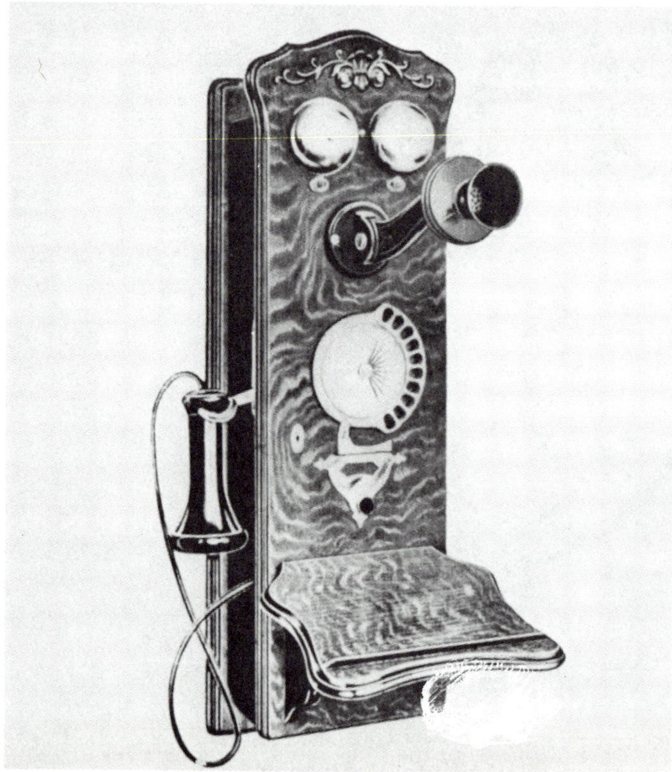

(Courtesy of Bell Laboratories)

operating in violation of the antitrust laws. In January 1913 the U.S. attorney general responded by advising AT&T that, in his opinion, some planned acquisitions of independent telephone companies were in violation of the Sherman Antitrust Act. That same month the Interstate Commerce Commission began an investigation to determine whether or not AT&T was monopolizing communications in the United States.

Faced with mounting public sentiment against monopolies and a probable government antitrust suit that could break up the company, AT&T reconsidered its position. In December 1913 AT&T vice-president Nathan C. Kingsbury sent a letter to the U.S. attorney general committing AT&T to:

1. disposing of its stock in Western Union
2. refraining from acquiring additional independent telephone companies except as approved by the Interstate Commerce Commission
3. interconnecting its facilities with those of the independents so that the independents could offer nationwide telephone service to their customers

This letter is generally referred to as the *Kingsbury Commitment.* The commitment was a victory for the independents because it prevented their being absorbed by the Bell System and assured them of interconnection with the Bell System. Also, like the Bell operating companies, the independents became monopolies within their respective areas. The commitment established the independents as an integral part of the communications industry and marked the beginning of the two groups' working together toward common goals.

The Graham Act

In 1921 Congress passed the Graham Act, which exempted telephony from the Sherman Antitrust Act in terms of consolidation of competing companies, thus legitimizing AT&T's monopoly.

Summary

The electric telegraph and a code system for its use were developed by Samuel F. B. Morse. After a demonstration of his invention to President Van Buren and his cabinet, Congress appropriated $30,000 to install a telegraph line between Washington, D.C. and Baltimore. On May 24, 1844, Morse sent the first public telegram over this line transmitting the message, "WHAT HATH GOD WROUGHT!"

Telegraph lines were strung on poles beside the railroad and by 1860 linked most of the major cities in the country. The telegraph was a vital contributor to the development of the railroads, the newspapers, and the New York Associated Press.

The Civil War years saw a dramatic rise in the use of telegraphic communications by businesses and the military. The network of railroads and telegraph lines in the northern states was an important advantage to the Union Army during the Civil War.

The early telegraph system was composed of many small companies. This meant that in order to send a long distance telegram, the message had to be retransmitted from company to company. One of the telegraph companies whose business prospered during the Civil War was Western Union. Gradually, Western Union bought out all of the competing telegraph companies, and by 1866 it was the nation's largest corporation and its first powerful monopoly.

The telephone was invented in 1876 by Alexander Graham Bell, a teacher of speech to the deaf. In 1877 the telephone industry was formally organized with the creation of the Bell Telephone Company. The first organization was a trusteeship. This organization made a decision that was to have far-reaching effects—to rent telephones rather than sell them. Selling telephone service only became a major practice of the telephone industry.

In 1877 Western Union organized the American Speaking Telephone company and began offering telephone service to the public. After a period of intense competition, Bell initiated a lawsuit against Western Union charging patent infringement. Convinced that it would lose if it pursued the matter in court, Western Union reached an out-of-court settlement with Bell providing that Western Union withdraw from the telephone business and sell its telephone network to Bell. Bell, in turn, agreed to stay out of the telegraph business.

In 1885 the American Telephone and Telegraph Company was formed to build and operate long lines for nationwide telephone service. In 1900 AT&T became the headquarters company of the Bell System. Before the expiration of the Bell patents, many new telephone companies were formed. Some of the new companies were organized under a Bell franchise agreement; others began operating "independent" of any Bell affiliation and in direct competition with the franchised companies. After the expiration of the Bell patents, competition between Bell and the independents intensified. Bell employed several tactics to combat the independents, including rate cutting, buying out competitors, and refusing to interconnect with Bell's lines.

In 1909 Bell bought controlling interest in Western Union and Bell's president, Theodore N. Vail, became president of both companies. In 1911 AT&T consolidated the franchise companies operating under Bell patents into state or territorial units.

In 1912 the independent telephone companies protested to the U.S. Department of Justice that AT&T was operating in violation of the antitrust laws. Faced with a probable lawsuit, AT&T vice-president Nathan Kingsbury sent a letter to the U.S. attorney general committing AT&T to disposing of its stock in Western Union, refraining from acquiring additional independent telephone companies, and interconnecting its facilities with those of the independents. This letter became known as the Kingsbury Commitment.

Review Questions

1. Why was the invention of the telegraph so important?
2. In Morse code, the letter *e* is represented by a single dot, and it requires the least electricity to transmit. Why?
3. Why was it disadvantageous to have many small telegraph companies?

4. How did the telegraph play an important part in the development of the railroads?
5. What was the impact of the Civil War on the telegraph industry? What was Anson Stager's role in uniting the telegraph industry?
6. What was the impact of the Civil War upon Western Union?
7. In the early days of the telephone, how was Western Union able to begin offering telephone service?
8. What did the Bell Company do to counter Western Union's telephone service?
9. What important principle was decided by the out-of-court settlement between Western Union and the Bell Company?
10. Why was it important for the Bell Company to have a single source of supply for telephone equipment?
11. Historically, what has been the role of AT&T in telephony?
12. What is the significance of the word *independent* in the independent telephone companies? What role did these companies play in early telephony?

References and Bibliography

Barret, R. T. *The Changing Years As Seen From the Switchboard.* New York: AT&T, 1936.

Bell Telephone System. *Alexander Graham Bell.* New York: Bell Telephone System, no date.

Boettinger, H. M. *The Telephone Book.* Croton-on-Hudson, N.Y.: Riverwood Publishers, Ltd., 1977.

Brooks, John. *Telephone.* New York: Harper & Row, 1976.

Fagen, M. D., ed. *History of Engineering and Science in the Bell System.* Bell Telephone Laboratories, Incorporated, 1975.

Harlow, Alvin B. *Brass-Pounders, Young Telegraphers of the Civil War.* Denver: Sage Books, 1962.

Jesperson, James, and Jane Fitz Randolph. *The Story of Telecommunications.* New York: Atheneum Publishers, Inc., 1980.

Pool, Ithiel de Sola, ed. *The Social Impact of the Telephone.* Cambridge, Mass.: MIT Press, 1977.

Rhodes, Frederick Leland. *Beginnings of Telephony.* New York: Harper & Brothers Publishers, 1929.

Shippen, Katherine B. "Mr. Bell Invents the Telephone." New York: Bell Telephone System, 1955.

Thompson, Robert Luther. *Wiring a Continent, the History of the Telegraph Industry in the United States 1832–1866.* Princeton, N.J.: Princeton University Press, 1947.

United States Independent Telephone Association. *Phone Facts '83.* Washington, D.C.: U.S. Independent Telephone Association, 1979.

———. "The Ring of Success." Washington, D.C.: U.S. Independent Telephone Association, 1979.

Watson, Thomas A. "The Birth and Babyhood of the Telephone." New York: AT&T, 1971.

Structure and Regulation of the Telecommunications Industry

3

The nation's telecommunications industry is currently undergoing its biggest reorganization since the "improvement in telegraphy" was patented in 1876. Except for a relatively brief period of competition in the first decade of the twentieth century, the telecommunications industry has been allowed to operate as a protected monopoly. Although in theory government regulation is designed to protect the consumer, in practice it has protected the monopolies and deterred technological advances. After over 70 years of permitting the telecommunications industry to maintain the comfortable status quo, the FCC changed its long-standing policies and embraced competition.

The transformation of the telecommunications industry from a monopoly system to a competitive system is being implemented in major part by breaking up the industry's dominant firm, AT&T. It also is being accomplished through the blurring of boundaries between data processing and telecommunications.

This chapter describes the telecommunications common carriers, the organizations that comprise the industry, the development of regulatory agencies, government regulation, the recent FCC rulings and court decisions that underlie the gargantuan reorganization, and the post-divestiture structure of the industry.

Common Carriers

In its broadest sense, the term *common carrier* describes any supplier in an industry that undertakes to "carry" goods, services, or people from one point to another for the public. In telecommunications, such "carriage" concerns provision of transmission capability over the telecommunications network. A common carrier that offers communications services to the public is subject to regulation by federal and state regulatory commissions.

Common Carrier Principle

The *common carrier principle* refers to the regulatory concept that limits the number of companies that can provide certain essential public services (such as utilities and transportation) in a specific geographic area. Its aim is to avoid unnecessary and expensive duplication of services and facilities, while providing strict standards of accountability to regulatory agencies for prices, return on investment, and services. This concept is designed to provide efficient utilization of telecommunications facilities, resulting in quality service at reasonable prices. The telecommunications common carriers offer facilities for the transmission of voice, data, facsimile, and television; some carriers also offer online computer services.

Historically, the giant American Telephone and Telegraph Company has been the largest of the domestic communications carriers. It consisted of 23 separate, but closely interrelated, companies. Other telecommunications common carriers include Western Union; General Telephone and Electronics; the independent telephone companies; specialized common carriers such as MCI, Southern Pacific Communications, Western Telecommunications, RCA, Graphnet Systems, Satellite Business Systems, and American Satellite; and value-added carriers such as Tymshare, University Computing Company, and Boeing Computer Services.

Companies Providing Common Carrier Services

The Bell Telephone System AT&T is the largest of the telecommunications common carriers. The following description portrays the historical composition of the Bell System prior to the Modified Final Judgment in 1982 (see Chapter 2). The present structure of the telecommunications industry will be discussed later in this chapter.

The *Bell System*, often referred to as "Ma Bell," consisted of the companies controlled by AT&T. The units that comprised the Bell System were AT&T; the 23 Bell Associated Companies, later known as the Bell operating companies (BOCs); Western Electric Company; and Bell Telephone Laboratories (Figure 3.1).

AT&T was the headquarters company of the Bell System. Subject to the regulation of the Federal Communications Commission, it interconnected the BOCs by means of its long distance lines; provided a centralized advisory service; controlled Western Electric, the manufacturing and supply unit for the system; and maintained Bell

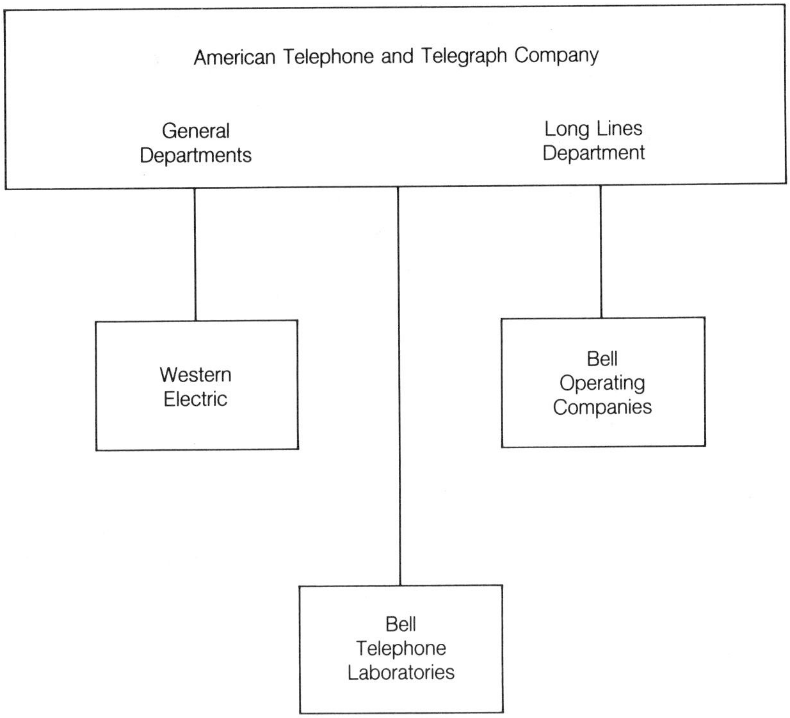

Figure 3.1
Historic Organization of the Bell System

Telephone Laboratories, Inc., an extensive organization devoted to research, development, and design in the communications field.

The services rendered by the parent company to the operating companies included manufacturing; financing; engineering; technical research; and advice and assistance on operating, legal, accounting, and financial matters.

To handle the general problems common to all the BOCs and to avoid duplication of expense and effort, the parent company provided the BOCs with services that could best be performed by a centralized organization. This was accomplished under a license contract that required the BOCs to pay a license fee to AT&T in return for these services.

AT&T and the 23 BOCS comprised the operating units of the Bell System. The BOCs were operated and managed by local personnel and were responsible for providing telephone service in the areas in which they were established. The long lines of AT&T linked each regional company with all the others; these lines extended the service given in each locality to every other part of the country. The BOCs operated under state laws and were subject to the regulation of the state or states in which they operated.

The BOCs were owned either wholly or in part by AT&T. They paid a license contract fee to AT&T as part of their operating costs. In addition, they paid all of their net profits to AT&T. These profits were the principal source of AT&T's revenue, enabling AT&T to pay dividends to its stockholders.

Western Union Western Union was the first telecommunications company in America. It was founded in 1856 to market the invention of telegraphy. The company grew rapidly and was soon operating a telegraph service to all parts of the United States.

In 1943, Western Union acquired Postal Telegraph, Inc., its most serious competitor. Following World War II, the market for telephone services rose sharply, while the demand for telegraph services declined substantially. Western Union strengthened its position by offering a teletypewriter service (Telex), and private-line services for voice, data, and facsimile. In 1971, Western Union purchased the Teletypewriter Exchange Service (TWX) from AT&T and began offering this service to its customers.

Western Union Telegraph Company is now a subsidiary of Western Union Corporation. In addition to its basic telegram service, it offers a money order service, public-switched teletype services (Telex and TWX), computer-switched teletype and data terminal services, and leases dedicated circuits to transmit voice, data, and facsimile messages. In cooperation with the United States Postal Service, Western Union provides a message service known as Mailgram.

Many of Western Union's services are provided by satellite. It has two WESTAR satellites and a wide range of channels that carry its own traffic and are also leased to other common carriers.

General Telephone and Electronics Corporation GTE services about 8 percent of American telephones, making it the second largest domestic communications common carrier and the largest independent telephone company. It offers a wide range of voice and data services. General Telephone and Electronics is commonly referred to as the General System.

General has two subsidiaries that manufacture telephone equipment both for itself and for sale to other companies: Automatic Electric and GTE Lenkurt, Incorporated. Other subsidiaries include Sylvania Electric, The British Columbia Telephone Company (Canada), and Telenet, a packet-switching network. In June 1983 GTE acquired the SPRINT network from Southern Pacific Communications.

The Independent Telephone Companies The non-Bell telephone companies are known as independent telephone companies. They

Year	Number of Independent Telephone Companies
1920	9,211
1970	1,841
1975	1,618
1980	1,483
1981	1,459
1982	1,432
1983	1,429

Figure 3.2
Decline of Independent Telephone Companies

serve about 20 percent of the telephones in the United States. The independent companies interconnect with long distance carriers to provide service to virtually anywhere. The Bell System is barred from acquiring independent companies by the Kingsbury Commitment.

Although some of the companies have grown rapidly, Figure 3.2 lists how the total number of independent companies has been steadily decreasing. This was caused by a trend toward the consolidation of smaller companies into larger independent companies.

Most of the independent telephone companies are members of the United States Telephone Association (USTA), formerly known as the United States Independent Telephone Association (USITA), which provides an organizational structure to represent them in matters concerning long distance revenues, technical standards, and regulatory matters.

Specialized Common Carriers The MCI decision opened the door to competition in the provision of long distance telephone service. This resulted in the development of a new type of common carrier, the specialized common carrier (SCC), which provides long distance service over microwave or leased-line facilities. Many of the SCCs' services are designed for customers who transmit large volumes of data. As common carriers, the SCCs are required to meet many of the same regulatory standards as the telephone companies; however, they are allowed to file for tariffs under different regulatory rules, which results in a different pricing structure.

The first SCC licensed by the FCC was Microwave Communications, Inc. (MCI). It began providing long distance telephone service between St. Louis and Chicago in 1972, relaying telephone calls over microwave stations. Initially, MCI served only major cities; however, it now provides service to every city in the United States. Microwave Communications, Inc., the original FCC petitioner, is the parent company of a number of affiliate companies, the largest of which is MCI Telecommunications Corporation, the long distance company. Although considerably smaller than AT&T, MCI is a leading competitor in providing long distance service in the United States.

The first SCC to offer coast-to-coast service was Southern Pacific Communications Company (SPC), which in 1974 began to operate its own private microwave relay system. Originally owned by Southern Pacific Transportation Company, its communications system was designed to support its transportation activities. Services provided to the public represented excess capacity in the system. In June 1983 Southern Pacific Communications Company and Southern Pacific Satellite Company were purchased by GTE. SPC's long distance service is called Sprint and is a major competitor of AT&T's Direct Distance Dialing Service.

Value-Added Carriers In 1973 another new type of common carrier known as a *value-added carrier (VAC)* emerged. A VAC leases transmission facilities from existing common carriers and adds computer-controlled services that increase the value of the basic transmission facility. Services provided by VACs include data transmission, facsimile, store-and-forward message systems, packet switching, electronic mail, and voice mail.

International Common Carriers International carriers provide a broad spectrum of telecommunications services between nations. The largest international carriers are AT&T Communications, International Telephone and Telegraph World Communications, Radio Corporation of America Global Communications, and Western Union International (owned by MCI). These carriers operate only from designated "gateway" cities, which include New York, Washington, Miami, New Orleans, Los Angeles, and San Francisco.

Government Regulation

In the years immediately prior to the Kingsbury Commitment (1913), the Bell System had adamantly refused to interconnect its long distance network with its competitors' lines. Many states viewed this as an abuse of power and concluded that it was contrary to the public interest. Accordingly, a number of states passed laws requiring the physical interconnection of telephone companies. It ultimately became apparent that a national regulatory agency was needed to regulate the communications industry more uniformly. The Mann-Elkins Act of 1910 gave the Interstate Commerce Commission (ICC) authority to regulate the financial practices of wire and radio communications. However, the ICC had been created to regulate interstate commerce and focused its attention on regulating transportation rather than communications.

Development of Regulatory Agencies

The first United States regulatory commissions were set up in the various states to regulate railroads. The federal government followed the states' examples with the creation of the ICC in 1887 and later with other commissions to exercise control over specific industries.

The first communications regulatory agencies were established in New York and Wisconsin in 1907; they were followed by similar agencies in all states. The FCC was created by Congress with the passage of the Communications Act of 1934. This act was the basis for communications regulation and created the national policy objectives of reasonable, universal service and the construction of the most rapid, efficient system possible.

Historically, the communications industry was considered a "natural monopoly" in which regulation would be in the "public interest" by preventing duplication of services. In practice, the most obvious function of regulation has been the monitoring and supervision of the firm's rate of return on its investment.

The Federal Communications Commission

The federal regulatory agency known as the Federal Communications Commission for many years consisted of a board of seven commissioners who were appointed by the president to staggered terms of seven years. In 1983 the membership was reduced to five commissioners. The law provides for bipartisanship by requiring that not more than a simple majority—three commissioners—may be from the same political party.

The FCC regulates interstate and international communications by radio, television, wire, and cable. Its responsibilities include encouraging the development and operation of broadcast and communications services at reasonable rates, regulating and licensing broadcast stations, reviewing and evaluating station performance, approving changes of ownership and major technical alterations, regulating cable television, prescribing and reviewing accounting practices, regulating and issuing licenses for all forms of two-way radio, reviewing applications of telephone and telegraph companies for changes in rates and service, setting permissible rates of return for communications common carriers, and reviewing technical specifications of new telecommunications equipment.

The *Office of Telecommunications Policy (OTP)* was created in 1970 by President Nixon to assist the government in formulating policies for the telecommunications industry. It reflected the growing importance of telecommunications in the United States and throughout the world.

The director served as the president's chief advisor on telecommunications. The office had the responsibility of ensuring that the views of the executive branch were effectively presented to the Congress and the FCC. The office had no executive authority; it could only issue recommendations.

In March 1978 the OTP was combined with the Office of Telecommunications of the Commerce Department to form the *National Telecommunications and Information Administration (NTIA).* Its function is to provide advisory assistance in telecommunications and information issues for the Department of Commerce. NTIA's broad goals include "formulating policies to support the development and growth of telecommunication industries; furthering the efficient development and use of telecommunications and information services; providing policy and management for federal use of the electromagnetic spectrum; and providing telecommunications facilities grants to public service users."[1]

Public Utility Commissions

Intrastate communications are regulated by the appropriate state Public Utility Commission (PUC). (In some states—Michigan, for example—the term Public Service Commission is used to identify the state regulatory agency.) PUCs exist in all states and the District of Columbia. Each telecommunications company must file a set of tariffs covering all standard intrastate service offerings with the regulatory body in its territory. This results in over 50 different sets of tariffs. Thus, there can be a wide difference in the rates of service from one state to another. Interstate tariffs, however, are uniform throughout the United States since they are under the jurisdiction of the FCC.

Although the format of state tariffs varies somewhat from state to state, the typical topics covered include:

1. general regulations (basic regulations covering use and provision of services)
2. local exchange service
3. general exchange service (includes key telephone, PBX, and other vertical services)
4. long distance message telecommunications service (intrastate)
5. Wide Area Telecommunications Service (WATS) (intrastate)
6. private-line service (intrastate)

1. *The United States Government Manual* (Washington, D.C.: United States Government Printing Office, 1983), p. 149.

Tariffs

The published rates, regulations, and descriptions governing the provision of communications services are known as *tariffs*. The tariff document

1. defines the services offered
2. establishes the rate the customer will pay for the service
3. states the general obligations of the public utility company and the customer in the provision and use of the service

All rates must be approved by the appropriate regulatory agency—federal or state—before they can become effective. Requests for rates for new services are submitted by filing new tariff schedules. Requests for rate increases or decreases are submitted by filing a petition. The commission can respond to such requests in any one of three ways: it can allow the schedule to take effect, it can reject the schedule, or it can delay the schedule and initiate public hearings on the request. When a public hearing is held, the public utility must present evidence to show why the petition's request should be granted.

Government Antitrust Suit of 1949 Against AT&T

In January 1949 the United States Department of Justice brought an antitrust suit against AT&T alleging violation of the Sherman Antitrust Act. The suit asked that Western Electric be separated from the Bell System and that it be split into three separate companies. The suit was settled in 1956 by a consent decree that permitted AT&T to retain its ownership of Western Electric but precluded Western Electric from manufacturing any equipment of a type not sold or leased to Bell System companies for use in furnishing common carrier communications services. It also limited the Bell System's activities to the telephone business and government projects. This meant that the Bell system would be barred from entering any type of electronic data processing or computer-related activity that was not limited to common carrier communications.

An Industry in Transition

For many years the BOCs and the independent telephone companies required their subscribers to use telephone instruments and other related communications hardware furnished by the telephone companies. Tariffs, filed by the telephone companies and approved by the regulatory agencies, specifically prohibited subscribers from attaching their own telephone equipment to telephone lines. The telephone companies claimed that since they were completely responsible for service quality, their equipment should be protected from any damage that might be caused by a device that the telephone company did not provide. Such equipment was referred to as a *foreign attachment.*

During the late 1950s and early 1960s many customers bought novelty and color telephones from other suppliers and connected them to the telephone lines without the company's knowledge or approval. If the telephone company were to discover the presence of a foreign attachment, the tariffs permitted them to disconnect the device or to terminate service. However, the enforcement of such tariffs proved to be an administrative impossibility.

The Carterfone Decision

In 1968 the Carter Electronics Corporation, a small Dallas-based manufacturing company, challenged the tariffs that prohibited foreign attachments. The proceeding began in 1966 as a private antitrust suit brought by Carter Electronics against AT&T and GTE in the United States Federal District Court in Texas.

Carter's original objective was quite limited. The company's primary interest was in marketing the Carterfone, an acoustic device used to interconnect private two-way radio communication systems with the telephone network. Carter sought to prove that the tariffs did not apply to the Carterfone since the device did not have an adverse effect upon the telephone system. The telephone companies argued that it did have such an effect.

In June 1968 the FCC decided that the Carterfone could be connected to the telephone system. The interstate tariffs were amended to provide that telephone companies be allowed to install a protective device to prevent damage to telephone company equipment. Thus, the FCC struck down existing interstate telephone tariffs prohibiting attachment or connection to the public telephone system of any equipment or device that was not supplied by telephone companies.

The Interconnect Industry

The Carterfone decision opened the way to competition in connection of customer-owned telecommunications equipment to the Bell and independent telephone company networks. The protective device portion of the tariff was later expanded to require approved FCC registration of any equipment to be connected to telephone company lines. As a result of the Carterfone decision, many companies started to build telecommunications equipment for sale to individuals and private businesses, and a new industry, which became known as the *interconnect industry*, was born.

Telecommunications equipment rapidly took on a new look, becoming decorative as well as functional. The variety in telecommunications devices and services appealed to subscribers, who enjoyed making their own selections. Businesses, too, liked to choose their own telephone systems. Many welcomed the opportunity to purchase telephone equipment outright instead of paying a continuing rental fee. A number of suppliers soon emerged to fill this new demand. A partial list of such suppliers includes ROLM, GTE, General Dynamics, Stromberg Carlson, MITEL, ITT, Northern Telecom, Rockwell International, Nippon Electronics, and OKI Electronics of America.

The MCI Decision

Shortly after the Carterfone decision opened the way for competition in interconnect hardware, the FCC decided that it would be in the public interest to allow "nontelephone" companies to provide common carrier services on a specialized basis in direct competition with existing common carriers. The decision was based on a case filed by Microwave Communications, Inc., requesting permission to provide intercity common carrier service by microwave for private, leased-line telephone use. The ruling, which was known as the MCI decision, also required that the existing telephone companies furnish interconnect services to the new common carriers. Because these common carriers specialized in a specific type of telecommunications service—intercity service—they became known as specialized common carriers (SCCs).

As a result of the MCI decision, other companies applied for and received permission to provide specialized telecommunications services. Originally limited to private-line long distance service, the SCCs later expanded their offerings to provide public-switched long distance service. These SCCs are also referred to as other common carriers (OCCs), a name given them by established common carriers when the SCCs first began competing for long distance business.

Specialized common carriers usually provide service to the high-density, low-cost intercity routes. They have constructed their own microwave and satellite facilities and, in addition, rent AT&T lines for resale. The two largest specialized common carriers are MCI and Southern Pacific Communications Corporation (SPC).

VACs provide services to customers who require more than basic transmission service. The added value concept is derived from the carrier's ability to store and forward data messages and to interconnect systems with different protocols. Packet switching is a service frequently offered by value-added carriers. The two largest VAC packet-switching operators are Telenet, owned by GTE, and Tymnet, a division of Tymshare.

FCC Computer Inquiry II Decision (1981)

As more and more computer companies became engaged in transmitting data over communications facilities, the question of whether they were engaging in common-carrier activities arose. As a result, the FCC investigated whether this activity should be subject to regulation. The FCC also considered whether or not AT&T should be permitted to process data during its transmission and thus engage in data processing activities. This investigation led to a decision known as the FCC Computer Inquiry II decision, which specified that:

1. computer companies be permitted to transmit data on an unregulated basis
2. the Bell System be permitted to engage in data processing activities
3. enhanced services and customer-premises equipment be deregulated and provided by a fully separated subsidiary of AT&T

As a result of this decision AT&T organized American Bell, a fully separated subsidiary and also renamed the Long Lines Department as AT&T Communications. Thus, the newly aligned AT&T in 1981 consisted of AT&T Communications, the long distance network; American Bell, the marketing arm; Western Electric Company; Bell Laboratories; and the BOCs.

Government Antitrust Suit of 1974 Against AT&T

In November 1974 the Justice Department filed an antitrust suit against AT&T, charging monopolization and conspiracy to monopolize the supply of telecommunications services and equipment and asking that Western Electric be separated from the Bell System. The suit also asked that some or all of the Long Lines Department and perhaps other parts of the Bell System be separated.

The suit was terminated in 1982 by a consent decree agreed to by the United States Justice Department and the AT&T. Later that same year the consent decree was modified slightly and approved by Judge Harold Greene of the United States District Court in Washington, D.C. It became known as the Modified Final Judgment (MFJ). Under terms of the MFJ, AT&T was required to divest itself of all its operating companies. Also, the Bell name was reserved for the exclusive use of the BOCs. As a result, American Bell changed its name to AT&T Information Systems

Figure 3.3
New Bell Operating Companies

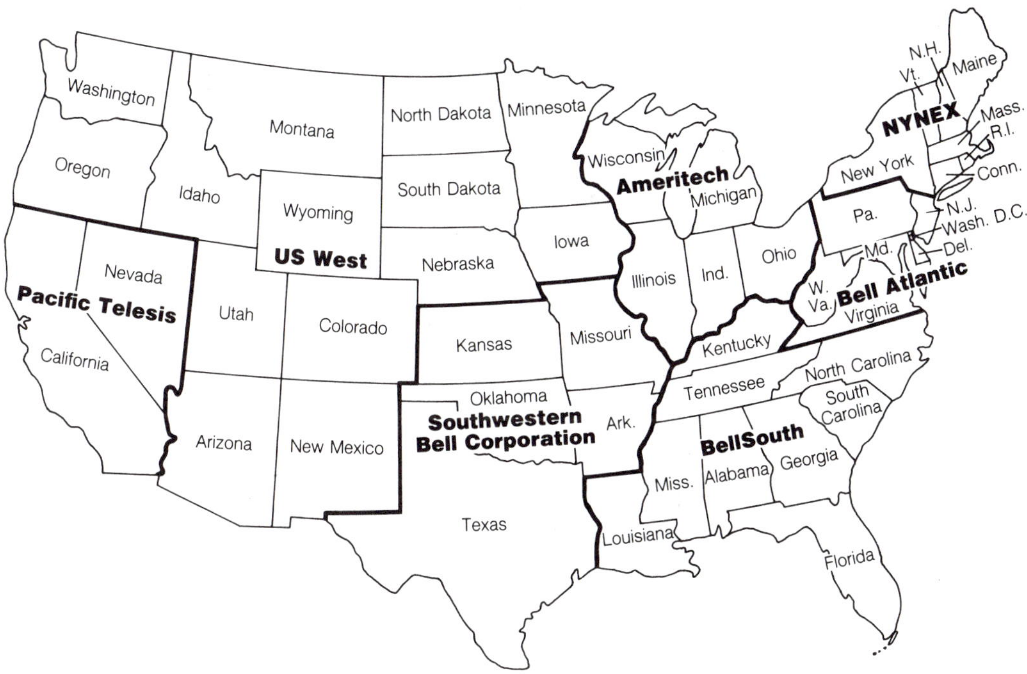

As a result of this divestiture, the spun-off BOCs were formed into seven regional corporations whose charter provides for monopoly carriage of local telephone traffic and its switching (Figure 3.3). The seven regions are NYNEX, Bell Atlantic, BellSouth, Ameritech, Southwestern Bell Corporation, US WEST, and Pacific Telesis Group. The regional companies were barred from providing customer-premises equipment. Local calling areas were mapped into 160 local access and transport areas (LATAs) throughout the United States. The operating companies were empowered to handle intra-LATA calls and to charge all long distance companies, including AT&T Communications, for connecting calls to and from their LATAs. Only long distance companies were empowered to provide telephone service between LATAs.

After divestiture, AT&T reorganized as AT&T Communications, the regulated long distance company (formerly Long Lines); and AT&T Technologies, a new unregulated corporation. The new corporation combines research, manufacturing, and business groups that market equipment and services. The groups in the new corporation include AT&T Information Systems (formerly American Bell), which operates under the conditions of the FCC's Computer Inquiry II ruling; Bell Laboratories; Network Systems; Consumer Products;

Figure 3.4
Post-Divestiture Organization of AT&T

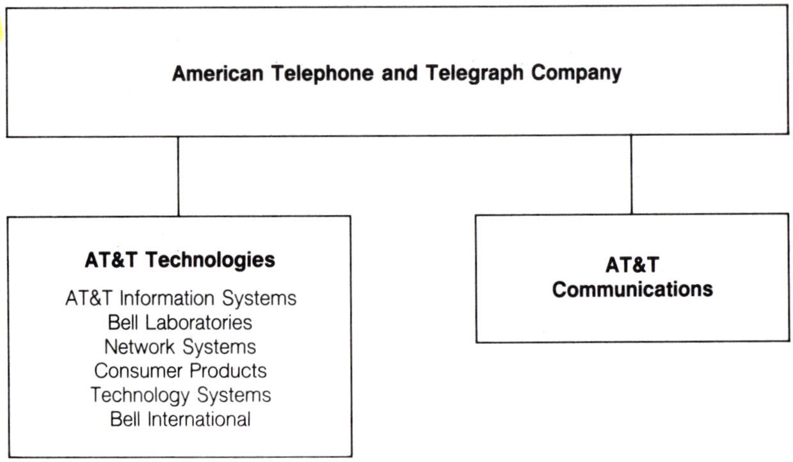

Technology Systems; and AT&T International (Figure 3.4). There is no longer a Western Electric Company, but the products produced by the various groups bear the Western Electric name.

Summary

Telecommunications common carriers provide the public with communications transmission services. They are regulated by federal and state regulatory commissions. The common carrier principle refers to the regulatory concept that limits the number of companies that provide essential public services in a specific geographic area. The telecommunications common carriers include AT&T, Western Union, General Telephone and Electronics, the independent telephone companies, the specialized common carriers, and the value-added carriers.

The first state communications regulatory commissions were established in 1907; in time, all states organized such commissions. The Federal Communications Commission was created by the Communications Act of 1934. Historically, the communications industry was considered a "natural monopoly" in which regulation was in the "public interest" by preventing duplication of services.

Tariff documents describe the published rates, regulations, and descriptions of telecommunications services. All rates must be approved by the appropriate regulatory agency before they can become effective. Interstate rates are regulated by the FCC; intrastate rates are regulated by state public service commissions. Intrastate tariffs vary widely from state to state; however, interstate tariffs are uniform throughout the United States.

The Bell System refers to the companies controlled by AT&T. Prior to the Modified Final Judgment, the units that comprised the Bell System were AT&T, the headquarters company; the 23 Bell operating companies; Western Electric, the manufacturing unit; and Bell Telephone Laboratories, the research and development organization.

In 1968 the Carterfone decision opened the way to competition in the supply of customer-owned telephone equipment connected to the telephone company networks. As a result, the interconnect industry was born. A few years later, the MCI decision opened the way to competition in long distance service. Because the new common carriers specialized in a specific type of telephone service, they became known as specialized common carriers. A type of specialized common carrier that provides a service over and above the transmission of voice or data is known as a value-added carrier. These added values are usually computer-oriented.

Two recent decisions caused a major restructuring of the telecommunications industry: the Computer Inquiry II decison (1981) and the Modified Final Judgment (1982).

The Computer Inquiry II decision specified that:

1. computer companies be permitted to transmit data on an unregulated basis
2. the Bell System be permitted to engage in data processing activities
3. customer-premises equipment be deregulated and provided by a fully separated subsidiary

The government antitrust suit of 1974 against AT&T was settled in 1982 by a Modified Final Judgment that required AT&T to divest itself of all its operating companies. It also prohibited AT&T from using the Bell name. The operating companies were formed into seven regional corporations to provide local telephone service and its switching. They were prohibited from providing or selling customer-premises equipment. The post-divestiture AT&T includes AT&T Communications (the long distance company) and AT&T Technologies. The latter organization consists of AT&T Information Systems, the marketing arm; Bell Laboratories; Network Systems; Consumer Products; Technology Systems; and AT&T International. There is no longer a Western Electric Company; however, the Western Electric name is used on the products produced by AT&T Technologies.

Review Questions

1. What is a telecommunications common carrier?
2. Name the principal companies that provide telecommunications common-carrier services.

3. What is the underlying principle of regulation?
4. What are some of the responsibilities of the FCC?
5. What is the principal function of the National Telecommunications and Information Administration?
6. What three functions does the tariff document serve?
7. Discuss the origin of the interconnect industry and describe its function.
8. Discuss the origin of the specialized common carriers. What type of telecommunications service do they provide?
9. Historically, what companies comprised the Bell System and what was the function of each company?
10. What was the principal thrust of the Computer Inquiry II decision?
11. What was the principal thrust of the Modified Final Judgment?
12. What companies comprise the post-divestiture AT&T and what is the function of each company?

References and Bibliography

AT&T Long Lines. *The World's Telephones.* Morris Plains, N.J.: AT&T Long Lines, 1981.

Blair, Roger D. "The Scope of Regulation in the Competitive Equipment Market." *Public Utilities Fortnightly,* December 17, 1981, 29–34.

Brock, Gerald W. *The Telecommunications Industry.* Cambridge, Mass.: Harvard University Press, 1981.

Brown, C.L. "AT&T and the Consent Decree." *Telecommunications Policy,* June 1983, 91–98.

Butler, Richard E. "The ITU's Role in World Telecom Development and Information Transfer." *Telephony,* August 23, 1983, 80–86.

Dizard, Wilson P., Jr. *The Coming Information Age.* New York: Longman Inc., 1982.

Fagen, M. D., ed. *Impact.* Murray Hill, N.J.: Bell Telephone Laboratories, Inc., 1974.

Flax, Steven. "The Orphan Called Baby Bell." *Fortune,* June 27, 1983, 87–88.

Fowler, Mark. "We're Heading Ultimately Toward a Regulation-Free Telecom Market." *Communications News,* March 1983, 100.

Geller, Henry. "Regulatory Policies for Electronic Media." *Telecommunications,* May 1983, 128–33.

Griesinger, Frank K. "Getting a Handle on the New Phone Rates." *Office Administration and Automation,* February 1984, 37–41, 87–90.

Iardella, Albert E., ed. *Western Electric and the Bell System.* New York: Western Electric Company, 1964.

Johnson, Ben and Sharon D'Amario Thomas. "Deregulation and Divestiture in a Changing Telecommunications Industry." *Public Utilities Fortnightly,* October 14, 1982, 17–22.

Jones, S. C. "Divestiture, Deregulation, and the Local Telco." *Telephony,* July 18, 1983, 160–66.

Kleinfield, Sonny. *The Biggest Company on Earth.* New York: Holt, Rinehart and Winston, 1981.

Kuehn, Richard A. *Interconnect: Why and How,* 2d ed. New York: Telecom Library, Inc., 1982.

Lewin, Leonard, ed. *Telecommunications in the United States: Trends and Policies.* Dedham, Mass.: Artech House, Inc., 1981.

Loomis, Carol J. "Valuing the Pieces of Eight." *Fortune,* June 27, 1983, 70–78.

Mabon, Prescott C. *Mission Communications, The Story of Bell Laboratories.* Murray Hill, N.J.: Bell Telephone Laboratories, Inc., 1976.

Martin, James T. *Telecommunications and the Computer,* 2d ed. Englewood Cliffs, N.J.: Prentice-Hall, Inc., 1977.

Mueser, Roland, ed. *Bell Laboratories Innovations in Telecommunications.* Murray Hill, N.J.: Bell Laboratories, Inc., 1979.

Pool, Ithiel de Sola, ed. *The Social Impact of the Telephone.* Cambridge, Mass.: MIT Press, 1977.

O'Reilly, Brian. "Ma Bell's Kids Fight for Position." *Fortune,* June 27, 1983, 62–68.

Uttal, Bro. "Western Electric's Cold New World." *Fortune,* June 27, 1983, 81–84.

Weber, Joseph H. "AT&T Restructure: 1982–1984, Its Causes and Effects." *Journal of Telecommunication Networks,* Spring 1983, 51–59.

4 Telephony

The telephone, born in America over a hundred years ago, has become the magic link by which a person can communicate with people across a street, across a city, or across a continent. The telephone is a wonderful device. It accepts the sounds of a human voice, transforms them into signals we cannot see or hear, and speeds them along the wires or through space to another telephone. There the sounds come forth, instantaneously delivering a replica of the original voice directly to the listener's ear. Throughout its history, the telephone has been an important force for human betterment. Today, telephones are so much a part of our lives that we take their presence for granted, using them without thought of the sophisticated technology and networks they use.

The telephone is the basic instrument of all communications technology. It provides the greater part of the world's communications business. The four primary types of telecommunications systems in use today are voice, data, message, and image. Voice communications, or *telephony*, refers to the electrical transmission of speech over a distance.

Principles of Telephony

The telephone works as a result of the application of a fundamental physical phenomenon: words spoken into a telephone mouthpiece are converted into electromagnetic impulses and transmitted over telephone lines. At the receiving instrument, these impulses are reconstructed into speech in such a manner that even the voice characteristics of the speaker are recognizable. This conversion phenomenon is basic not only to telephony but to the entire field of telecommunications as well.

The Telephone Instrument

All telephones consist of at least three parts: the transmitter, the receiver, and the bell unit. The transmitter is similar to a microphone; it converts voice vibrations into electrical impulses, which are then transmitted over telephone wires, radio waves, or satellites. The telephone receiver converts the incoming electrical impulses to sounds, and the bell unit rings or buzzes when activated by an incoming call.

Preautomatic telephone instruments were made up solely of these three parts and required the assistance of an operator to control the completion of the call. Today's modern instruments contain a *control unit*—either a rotary dial or a set of pushbutton keys—that the caller uses to place calls directly, without the assistance of an operator. An auxiliary part of the modern instrument is the *switchhook*, which signals the telephone company central office and incoming callers that the telephone is either idle or in use. The transmitter and the receiver are generally housed in the telephone's handset; the bell unit and the switchhook are located in its base. When not in use, the handset rests in such a manner that the switchhook is in the "off" position; when a user picks up the handset, the switchhook is released to the "on" position.

Telephone Lines

Each telephone line consists of a pair of wires. Early telephones were equipped with batteries so that they could provide their own electrical current; the telephone lines contained no flow of electrical current except when they were in use. Each telephone also contained a hand-operated electrical generator called a *magneto*. By cranking the handle on this generator, the user activated a bell or light that signaled an operator. When the operator answered the signal and manually connected the caller's line to the line being called, the electrical circuit was completed, current flowed, and the caller could talk to the party called. Modern telephone lines employ *common-battery operation*, in which a central source of electricity is always available for use.

Telephone lines connect individual instruments to a central point, known as a *central office*. Lines that connected a customer's telephone to a central office operator originally consisted of a single wire, over which current flowed in one direction; the ground served as a means of completing the electrical circuit. Because these single-wire lines provided extremely poor transmission, they were soon replaced with two-wire lines that formed a "loop" between the telephone and the central office. The wires were strung through the air on poles with crossarms to hold them apart. As the number of telephones increased, the poles and wires became so numerous and unsightly that their continued use was impractical, and several

Figure 4.1
Broadway and Maiden Lane, New York City

(Reproduced with permission of AT&T)

techniques were developed to compact the lines by placing them in pipes or cables. (See Figure 4.1.) Today, wires are packed into cables. Each wire is insulated from the other wires and covered with a lead or plastic sheath that is sealed to prevent water from entering the cable. The cables may be placed either overhead or underground.

The modern loop between telephones and central offices consists of a *drop wire,* a *distribution cable,* and a *feeder cable.* Each loop consists of a pair of wires. The drop wire runs from a residence or building to a telephone pole, where it is connected to a distribution cable. Several distribution cables are brought together at a junction and connected to a much larger feeder cable, which extends to the central office. Frequently the feeder cable is placed in underground ducts, reducing the need for unsightly telephone poles.

A second type of telephone line connects central offices to each other. These lines, known as *trunks,* are usually made up of wires in large cables that are almost always carried underground in cable ducts. Both kinds of lines form a network.

Central Offices

Bell's experiments were conducted at the electrical shop of Charles Williams, Jr., in Boston. Appropriately, when the first permanent outdoor telephone line was strung on April 4, 1877, it was between the electrical shop and Williams's home, a distance of about three miles. Shortly thereafter, the first telephones rented from Bell for business use were connected by a line from the Boston office of a young banker to his home in Somerville. These first telephone lines were strung directly from one individual telephone to another; no "central" point intervened.

As more telephones came into use and more and more wires were strung, it became apparent that stringing direct lines between separate instruments was no longer feasible. Moreover, people realized that the value of the telephone would be substantially increased if it were connected to more than one other telephone. Thus, the process of transferring a connection from one telephone circuit to another by interconnecting the two circuits was developed. This process is known as *switching*. The central office, or *telephone exchange* as it is sometimes called, came into being as a switching center—a point at which two circuits could be interconnected to make a talking path between two telephones. Originally, "Central" was used to identify the operator who manually connected telephone lines to each other. In time, automatic switching equipment performed this function, but the term *central office* is still used today to describe the place where call switching is done. Besides acting as a focal point of many telephone lines, the central office provides a common source of electricity to a number of telephones.

Each central office serves a specific geographic area, functioning as a switching center for the individual telephone lines that terminate there. The size of the area served depends on the number of telephone lines; large metropolitan areas require many central offices. Each office contains trunks to other central offices and to long distance switching offices, so that any telephone line can be connected to any other telephone line anywhere. These connections are implemented by the *central office switching equipment*, the general industry name for the mechanical, electromechanical, or electronic equipment that routes a call to its ultimate destination. During the routing process the call may be switched once or any number of times. The greater the distance a call has to travel, the more likely it is to require multiple switchings.

Each central office has the capacity to handle a certain number of calls at one time. If too many calls come into the central office switching center simultaneously, the system becomes overloaded and the customer either fails to get a dial tone or gets a fast busy signal before the dialing has been completed. This condition is known as *blocking*; it will occur occasionally because telephone companies have found that it is not economically possible to provide each central office with enough equipment to meet every calling load and condition during peak times.

Figure 4.2
Early Telephone Switchboard

(Reproduced with permission of AT&T)

Automatic Switching

In the early days when all switching was done manually, operators interconnected lines by inserting a cord with jack plugs into each of the two lines to be connected (Figure 4.2). As the use of telephones spread, manual call switching quickly became obsolete, partly because of the ever-increasing numbers of operators that would have been required. Thus, in the 1920s, manual switching began to be replaced by automatic exchanges, in which the caller dials the desired number directly to effect the connection between the two telephones. The equipment that makes the connections consists of banks of relays and switches mounted on rows of frames that extend from floor to ceiling and occupy entire buildings.

The story of the invention of automatic switching by a Kansas City undertaker, Almon B. Strowger, provides an interesting sidelight in the history of the telephone. According to tradition, Strowger suspected the local telephone operator of misdirecting calls intended for him to his competitor, who was a relative of the operator. Accordingly, he vowed to find a way to eliminate the need for operators

(Reproduced with permission of AT&T)

Figure 4.3
Step-by-Step Switching Equipment

and went to work to develop an automatic switching device that would make his competitor's accomplice obsolete. In 1891 a patent was issued to Strowger for such an invention: a two-motion (vertical and horizontal) switch that was subsequently named the *Strowger switch*. The principle on which Strowger's invention was based was so fundamental to automatic telephone systems that for many years telephone companies had to obtain a license from him before updating their equipment.

A central office switching system receives dial pulses generated by a calling telephone, registers the pulses in an electrical buffer, and translates them into a series of equipment operations that complete the call. The earliest automatic central offices were equipped with *step-by-step* systems (Figure 4.3), which offered a means of selecting a series of paths through which the circuit from the calling party to the called party could be progressively established. Each digit dialed activated one switch in the central office, and the sequence of digits selected the called telephone number for each connection one digit at a time.

Another early type of electromechanical switching equipment, the *panel system*, derived its name from the way groups of numbers were arranged on frames resembling panels. In this system all the digits dialed were stored in a control device and the call advanced at the completion of dialing, based upon the predetermined logic wired into the equipment.

Both step-by-step and panel systems were limited to 10,000 telephone lines. Although extremely reliable, these systems were slow

Figure 4.4
Crossbar Switching Equipment

(Reproduced with permission of AT&T)

by today's standards, because of their mechanical nature. In addition, maintenance proved costly because the mechanical equipment required frequent cleaning, lubrication, and adjustment. A subsequent development, the *crossbar system*, which derived its name from the way connections were established, consists of series of switches with vertical and horizontal *talking paths*. (See Figure 4.4.) At the intersecting points between these paths, or crosspoints, contacts are made on platinum or another precious metal that produces virtually noise-free connections. An important feature of crossbar equipment is its pretesting of trunk and line availability before it establishes connections. Moreover, the circuit connections in this type of equipment are established with fewer mechanical movements than a step-by-step switch, resulting in less wear and easier maintenance. Crossbar systems have a capacity of up to 30,000 telephone lines or three number groups of 10,000 lines each. (See Figure 4.5.)

Although both step-by-step and crossbar equipment can still be found in local exchanges and private branch exchanges throughout

(Reproduced with permission of AT&T)

Figure 4.5
Crossbar Switching Equipment

the United States, these traditional systems are increasingly being replaced by electronic switching systems that establish connections at phenomenal speeds and have a capacity of up to 100,000 lines (Figure 4.6). Electronic switching equipment is called *stored program control* switching because it can be programmed to perform a variety of functions in addition to conventional call completion. Although relatively expensive to purchase, it is economically attractive because of its low maintenance costs and revenue-producing service features.

Transmission of Sound

Simply stated, a telephone works as follows: sound waves generated by speech or other sources of sound are converted to electrical energy at a transmitting instrument and are sent over wires to a receiving instrument, where they are reconstructed as sound. The voice passes over the wire in the form of an *analog signal*, that is, in a continuous flow of electrical energy that oscillates in proportion to the frequency and intensity of the sound being transmitted. Sound that can be heard by the human ear consists of frequency ranges between about 30 Hz (Hertz) and 15,000 Hz. (*Hertz* is a standard unit of frequency that has replaced, and is identical to, the older unit of measurement called "cycles-per-second.") Telephone transmission equipment is designed to use frequencies varying from 300 Hz to 3,400 Hz—a

Figure 4.6
Electronic Switching System Equipment

(Reproduced with permission of AT&T)

range that is satisfactory for telephone transmission because it produces voice characteristics that are both intelligible and recognizable.

Analog vs. Digital Transmission

Digital transmission of sound uses a different physical process than analog transmission. In digital systems a stream of discrete "on" and "off" pulses, called *bits*, are sent over the transmission facility. Since analog signals can be converted to digital signals and digital signals converted to analog, any type of information can be transmitted in either analog or digital form. The conversion from analog to digital and/or from digital to analog is performed by a unit of equipment called a *modem*, which performs the functions of a modulator and demodulator. The modulator is used in transmission, and the demodulator in reception. The Bell System term *data set* is synonymous with the term *modem*, which is used industrywide.

The traditional telephone line consists of one pair of wires and can carry one analog signal—a continuous stream of frequencies. A trunk line, or talking path between two central offices, may consist of one pair of wires carrying one analog signal at a time. Trunk lines can also be constructed to carry digital signals by providing special equipment that breaks down the analog signal into discrete bits or pulses through a technique known as *pulse code modulation (PCM)*. By converting the analog signals to digital signals, the unwanted noise that is characteristic of analog transmission is virtually eliminated. The result is improved sound quality, because signal strength

Figure 4.7
Numbering Plan for North America

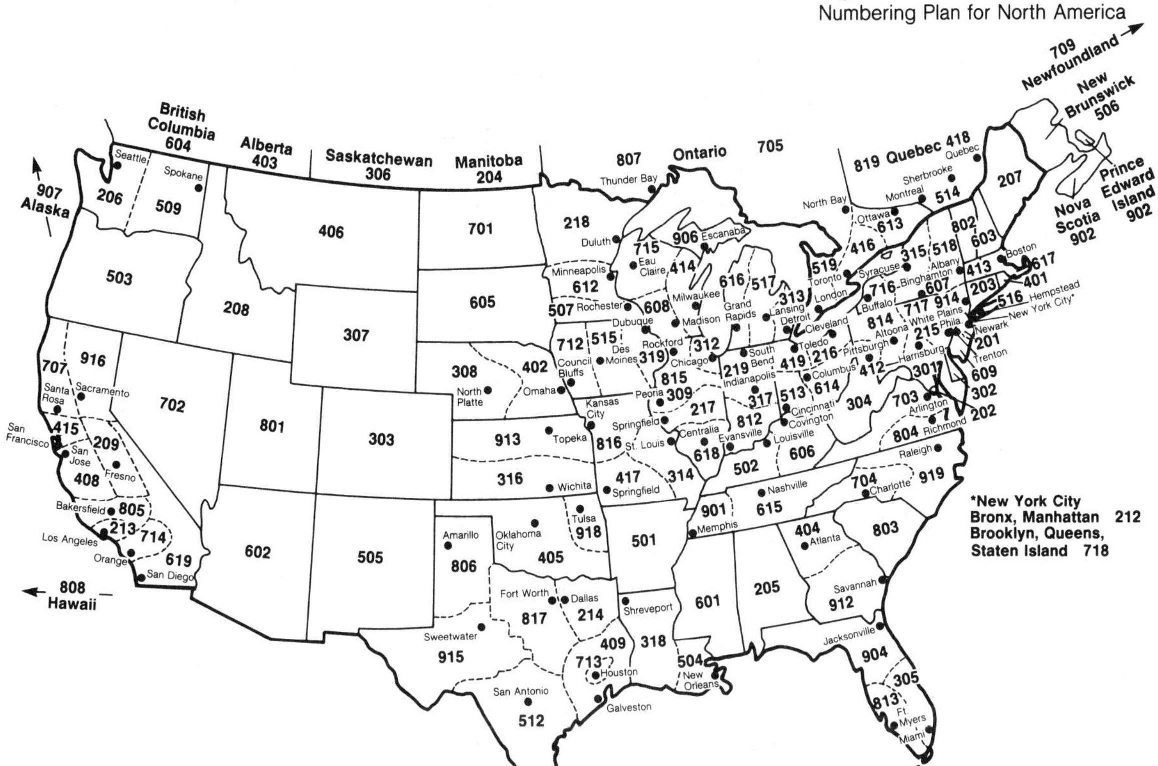

can be amplified without amplifying the noise and distortion. Digital transmission is also extremely efficient because the discrete bits can be transmitted at exceptionally high speeds. Because of their many advantages, digital switching and transmission systems are gradually replacing analog systems.

Nationwide Numbering Plan

The telephone industry has divided the country into geographic areas called *numbering plan areas (NPAs)*. Each NPA is identified by a unique three-digit code, called the *area code*, which precedes the local, seven-digit telephone number. (See Figure 4.7.) Thus, seven digits are required to place a call to any telephone within the same NPA, and ten digits are required if the telephone call is to terminate at a telephone in a different NPA. Because the digit "0" is reserved for operator calls and the digit "1" is reserved for identifying long distance calls, only the digits "2" through "9" are available for use as the first digit of a seven-digit telephone number. Mathematically, then, there can be no more than 8 million telephone numbers in a given geographical area.

Most telephone companies require that customers prefix the digit "0" on calls that require the assistance of an operator. The digit "1" is frequently called a *long distance access code*. Dialing the extra "1" helps remind the user that there will be a charge for the call and improves the credibility of the telephone bill. Access codes other than the "1" are generally not assigned in public telephone systems because even the use of a single digit as an access code would invalidate an entire block of one million digits for general telephone number use. Private telephone systems employ access codes such as the digit "9" to reach the public network, and other codes to meet the specific design of the private system. The use of these access codes does not interfere with the public telephone system and is compatible with the overall numbering plan.

World Numbering Plan

The International Telecommunication Union (ITU), an agency of the United Nations, was established to provide worldwide standards for communications. The union's Consultative Committee on International Telegraphy and Telephony (CCITT) is the medium for recommendations for international communication systems. By following CCITT recommendations, the United States and other United Nations countries achieve compatability between their telecommunication systems and make international communications possible. One of the CCITT's most important recommendations concerned a world-wide numbering plan that assigns each customer a unique international telephone number.

The plan assumes "all number calling," that is, telephone numbers composed entirely of numbers, with no letters of the alphabet being used. Under the plan, a customer's international telephone number consists of a maximum of twelve digits, plus access code (if one is required). The first one, two, or three digits represent the code assigned to the country, and the remainder constitute the national telephone number, which is composed of the routing code (area code) plus the subscriber's local telephone number.

In order to assign country codes, the world was divided into nine geographic zones, sometimes referred to as World Zones, each of which was assigned a digit, as follows:

1 North America
2 Africa
3 Europe
4 Europe
5 South and Central America
6 South Pacific

7 Union of Soviet Socialist Republics
8 Far East
9 Middle East and Southeast Asia

All countries within a zone were assigned codes beginning with that zone's digit. Since the plan was formulated to meet the telephone requirements of each country for the year 2000, the number of digits in a country code is determined by the number of national telephones forecast for that year. Countries with the highest requirements were assigned a single-digit number, and those with the lowest requirements were assigned a three-digit country code.

Because the composition of national telephone numbers is determined by each individual country, there are many variations in national numbering plans. As discussed in the previous section, United States national telephone numbers are composed of ten digits. By prefixing the digit "1"—the digit assigned to North America—the national number becomes the American's unique international telephone number.

International Dialing

International dialing is not universally available presently. When this service is available, the customer can place an international call by dialing an *international access code* followed by the *country code* plus a *routing code* (area code) plus the subscriber's telephone number. Thus, a call from the United States to Frankfort, Germany, would be completed by dialing 011-49-611-432-432 (011 is the international access code; 49 is the country code; 611 is the area code; and 432-432 is the subscriber's telephone number).

Customer Equipment and Services

A host of innovative equipment and service features are available to enhance the usefulness of the telephone. In the present post-divestiture environment, telephone companies provide switching services, lines, and trunks to complete local and long distance calls. Other vendors provide the CPE, such as telephone sets, PBXs, and ancillary devices, required to enable users to place and receive telephone calls.

Private Branch Exchange

A *private branch exchange (PBX)* is a switching system installed for the exclusive use of one organization; a PBX is usually located on the customer's premises. The system can place calls to or receive calls from the public network, as well as handle calls within the organization. Calls between stations (telephones served by the PBX) are dialed directly. One directory number is listed for the entire organization. Direct inward dialing (DID), or calls direct to certain extensions without use of a switchboard, can be provided by some PBXs.

The systems can be purchased, leased, or rented from the local telephone company or from any one of a number of other vendors. Justification for installing such a system depends on the number of telephones an organization needs and the ratio of internal telephone calls to external telephone calls.

Although a few manually switched PBXs remain in use today, most are completely automatic. The automatic PBXs are known as PABXs (*private automatic branch exchanges*); however the term PBX is sometimes used to mean any type of PBX, including one that is completely automatic. Some vendors refer to their computerized PABXs as CBXs (computerized branch exchanges).

Centrex is a type of PBX in which incoming calls can be dialed direct without the assistance of an operator. Outgoing calls are dialed direct by extension users. Centrex is a Bell System service in which the switching equipment is usually located on the telephone company premises, and each telephone served by the system is directly connected to the telephone company switching center. This arrangement differs from the usual PBX wherein the switching equipment is generally installed on the customer's premises.

Computerized PABXs offer many service features not available on earlier types of PABXs. (See Figure 4.8.) When these systems were first introduced, their cost put them beyond the reach of most organizations. However, in the last few years the prices of small computers have decreased considerably, with a resulting decline in the cost of computerized PABXs. In addition, some operational features can control telephone costs and make computerized PABXs more economical.

Computerized PABXs are available from most vendors of telephone equipment. The heart of these systems is typically a small computer with a solid-state switching network. Frequently used programs are stored in the computer's memory; new features can be added by updating the computer software. Although each product's features differ somewhat from that of other companies, a number are found on nearly all systems. Some of the available features include:

- ☐ *Automatic call forwarding.* By dialing instructions to the computer, a person may direct the PABX to forward incoming calls to another location.

(Courtesy of ROLM™)

Figure 4.8
Computerized PABX

- *Automatic call stacking.* When calls arrive for a station that is busy, they can be automatically answered and a recorded "wait" message played to the caller.
- *Automatic call back.* When a call is placed to a busy number, the caller may instruct the PABX to call back when the number is free.
- *Telephone number retention.* Users who move can retain the same telephone numbers when they move to a new station because the tables used by the PABX are easily changed; no rewiring is necessary.
- *Pushbutton station selection.* The attendant has a status light to indicate when the line is busy and a button for controlling each station.
- *Call waiting ("camp-on") signal.* A person using a line can be signaled to indicate that another call is waiting.
- *Outgoing call restriction.* Designated stations may be prevented from making outgoing calls.
- *Call holding.* A user can place a call on "hold" while dialing or talking to another station, then return to the interrupted call.
- *Third party add-on.* A user can dial another station, thereby setting up a three-party conference call.

- ☐ *Charge listing by stations.* The computer can furnish a listing of the toll calls placed by each station, giving the number, time, duration, and cost of each call.
- ☐ *Pushbutton to dial pulse conversion.* Signals can be converted so that pushbutton telephones can be used even when the local central office accepts only rotary-dial pulsing.
- ☐ *Abbreviated dialing.* Commonly used telephone numbers are replaced by two-digit numbers to shorten dialing time.
- ☐ *Automatic call transfer.* Incoming calls to a busy station can be automatically transferred to another designated station.
- ☐ *Distinctive ringing.* Different ringing tones are used to identify incoming calls so that the person called has some indication of the call's origin before he or she answers. Ringing tones may distinguish between internal, external, secretary, and specific extension calls.
- ☐ *Automatic least-cost routing on corporate networks.* A call is routed from the corporate network by the least expensive route; for example, first choice, tie line; second choice, WATS line; third choice, direct distance dialing.

Non-PBX Telephone Services

The largest segment of telephone service is provided for residences and businesses not served by a PBX. These customers require single- or multiple-line telephone instruments. Key telephone sets permit more than one line to be terminated on one telephone instrument; any one of the lines can be used by pressing the key associated with that particular line. Key sets permit users to place calls on "hold" while they switch to another telephone line. Also, these instruments are often equipped with "interoffice" features that allow users on various lines to talk to each other directly. Some of the features previously listed for PBX customers are also available on single- or multiple-line telephones. Their availability is determined by the capability of the central office. Service features most frequently available include call forwarding, call waiting, third party add-on, and abbreviated dialing.

Today's Wide Range of Telephones and Special Features

Many different kinds of telephones are now offered to customers and can be obtained from phone centers, department stores, and electronics stores. The telephones are equipped with a connector plug that fits into a standard jack or outlet, enabling the customer to attach them to a house or building's internal wiring without the assistance of a telephone installer. In addition, many telephone instruments are

Figure 4.9
TOUCH-TONE® Desk Type Telephone Instrument

(Reproduced with permission of AT&T)

modular; they can easily be taken apart and reassembled. Modular construction facilitates the replacement of defective components in a telephone, reducing telephone maintenance costs.

Until the middle 1950s, telephones came in only one color—standard black—but today all telephone manufacturers offer a wide variety of decorator colors to harmonize with many decors. Telephones are also available in a variety of shapes and designs: replicas of antique telephones, reproductions of storybook characters, and novelty items of many types and descriptions.

Rotary-dial telephones are gradually being replaced by pushbutton sets (Figure 4.9). Each pushbutton controls the transmission of a pair of frequencies that produces a tone resembling a musical note. Each digit has its own distinctive tone that can be identified by the receiving equipment. Pushbutton telephones have several advantages over rotary-dial telephones. Some examples are:

1. Telephone numbers can be keyed much more quickly than they can be dialed, because the dial return interval is eliminated.
2. Impatient users of rotary-dial telephones frequently reach wrong numbers because they often try to speed up the dialing process by forcing the dial mechanism back to its original position.

Figure 4.10
Telephone Answer and Record Unit

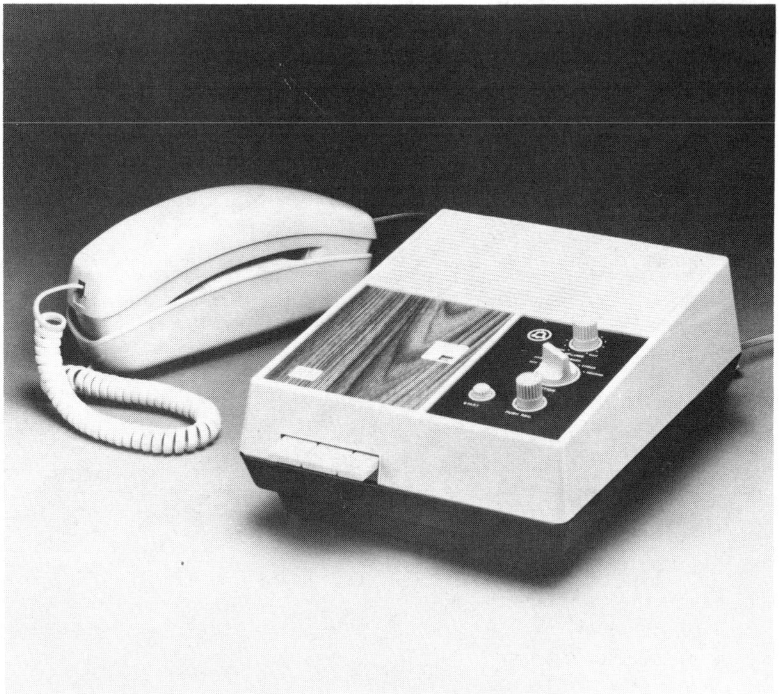

(Reproduced with permission of AT&T)

3. Because of the shorter dialing time, the pushbutton caller does not have to remember the number being called for as long a time, which helps dialing accuracy.

Many varieties of key telephone sets permit several lines to be terminated on one telephone instrument. Most vendors offer top-of-the-line key sets that can be customized to provide additional service features similar to those provided by PBXs; they are known as "hybrid" systems.

Many specialized telephone features have been designed to meet a wide range of customer needs and to make communication increasingly convenient. (See Figures 4.10 and 4.11.) Familiar examples are the distinctive hotel and motel systems that can light a "message waiting" lamp or restrict incoming calls if a client doesn't want to be disturbed. These features, offered by all telephone vendors, also provide one- or two-digit dialing for room service and other hotel services. A newer innovation, the "Touch-a-matic" feature (Figure 4.12), permits the placing of a call by name or number to 33 selected telephones by pressing one key; a "last call placed" key also enables a user to re-call a telephone number that previously was busy or did not answer.

Figure 4.11
Speakerphone

(Reproduced with permission of AT&T)

Also available are cordless, remote radio-controlled telephones that enable a user to conduct a telephone conversation at a short distance from the regular telephone instrument through the use of a radio-controlled talking path.

Another specialized service, "speakerphone," consists of a transistorized, voice-switched, microphone-speaker system that permits hands-free conversation. The user can listen and talk without holding a handset and can move about the room while continuing the conversation. This feature is sometimes combined with *teleconferencing,* an increasingly popular service that involves equipping meeting rooms with microphones and speakers so that a group of conferees can communicate with other people in similarly equipped rooms in other locations (Figures 4.13–4.15). A related development, the Bell System's "Gemini Blackboard," permits each conference room to use an easel to produce outlines or sketches that are duplicated on a screen in the other conference room. Many businesses are finding it economically attractive to substitute teleconferencing for personal attendance at conferences, thereby reducing or eliminating the need for expensive, time-consuming travel.

Conference rooms may also be equipped with television cameras, screens, and telephone circuits to provide picture telephone service.

Figure 4.12
Speed Dialer. The Bell System's Touch-a-matic® "S" telephone incorporates a dialing device run by a microprocessor that direct dials any one of 12 preprogrammed telephone numbers with the touch of a single key.

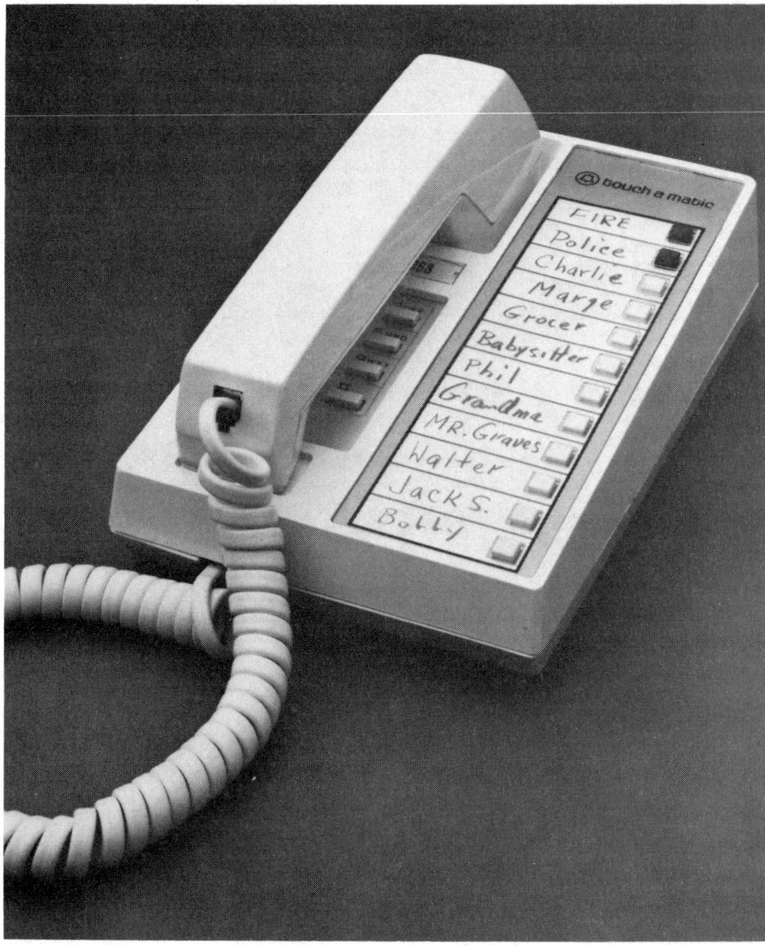

(Reproduced with permission of AT&T)

When two or more conference rooms are so equipped and connected, two-way conferences can take place. In some of these systems each conferee is provided with a microphone that controls a voice-activated switch. When a conferee speaks, the switch activates the camera that automatically focuses on the person speaking. Because the cost of this type of service is substantially greater than that of teleconferencing, the service is not widely used at this time.

In today's mobile society there is great demand for telephone service between stationary telephones and moving vehicles. *Mobile telephone service* employs both the telephone network and a radio circuit to establish communication. The stationary telephone is connected to a fixed antenna through the telephone network; a radio circuit provides the connection from the antenna to the vehicle.

(Reproduced with permission of AT&T)

Figure 4.13
The Bell System's QUORUM™ Teleconferencing Microphone and Loudspeaker with a vertical design that houses a series of sensitive microphones that reduce echo and the transmission of extraneous noises. It can free teleconference participants from the constraints of conventional microphones.

Summary

Telephony refers to the electrical transmission of speech over a distance. The telephone is the basic instrument of all communications technology.

The telephone works as a result of the application of a fundamental physical phenomenon: words spoken into a telephone mouthpiece are converted into electromagnetic impulses and transmitted over telephone lines. At the receiving instrument, these impulses are reconstructed into speech so that even the voice characteristics of the speaker are recognizable. This phenomenon is basic not only to telephony but to the entire field of telecommunications as well.

Telephone systems consist of four basic parts:

1. telephone instruments
2. telephone lines that connect the telephone instruments to a central office
3. the central office that interconnects the telephone lines
4. trunks that interconnect central offices so that any telephone caller can reach any other telephone

Figure 4.14
The QUORUM™ Teleconferencing Bridge. It connects locations for a teleconference and allows up to 28 locations to meet electronically.

(Reproduced with permission of AT&T)

The process of transferring a connection from one telephone circuit to another by interconnecting the two circuits is known as switching. The central office serves as a switching center. In the early days of the telephone, all switching was done manually. Operators interconnected lines by inserting a cord with jack plugs into each of the two lines. In the 1920s manual telephones began to be replaced by automatic exchanges in which the caller dials the desired number directly to make the connection between two telephones.

In order to connect any telephone to any other telephone, it is necessary to have a numbering plan that identifies each telephone as unique. In the United States, each telephone subscriber is assigned a seven-digit telephone number, which is used for calls within a specified local area. Additionally, the nation is divided into geographical areas, each of which is assigned a three-digit code, called the area code, which precedes the local telephone number. International calls are dialed by prefixing a country code to the national telephone number of the called telephone. Further, world zones have been created, so each country is assigned a country code beginning

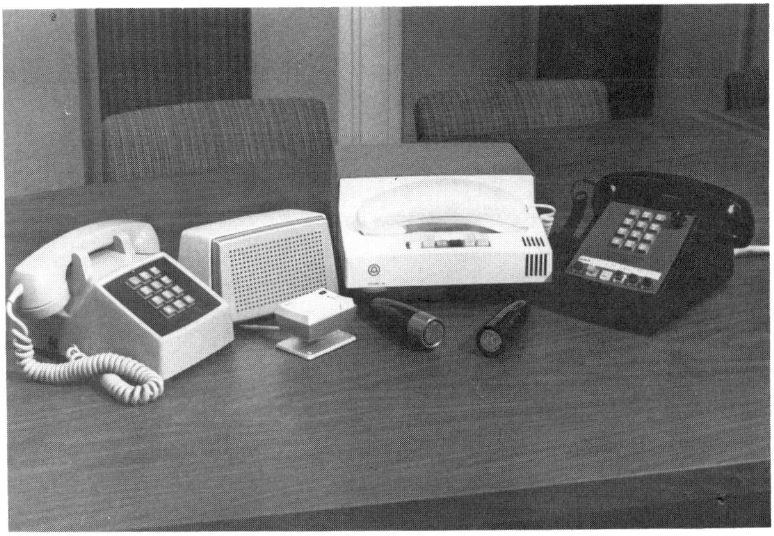

(Reproduced with permission of AT&T)

Figure 4.15
The QUORUM™ Teleconferencing System. It includes a variety of voice terminals used to conduct the audio portion of a teleconference. Shown, from left to right, are the Speakerphone, the QUORUM Portable Conference Telephone, and the control unit for the QUORUM group Audio Teleconferencing Terminal.

with that zone's digit. Generally, an access code is also required to reach the international network.

Private branch exchanges are switching centers installed for the exclusive use of one organization. Although these PBXs can place calls to or from the public network, their main function is to handle calls within the organization. Most PBXs in use today are completely automatic. Computerized PBXs (known as PABXs or CBXs) contain a small computer with a solid-state switching network. Computerized PBXs are capable of providing a variety of service features such as call forwarding, call stacking, automatic call back, telephone number retention, call waiting, outgoing call restriction, call holding, third-party add-on, charge listing by stations, abbreviated dialing, call transfer, distinctive ringing, and least-cost routing.

Commercial customers whose needs do not justify PBX service may use key telephone sets, which allow more than one telephone line to be terminated on a single instrument. Most key telephone sets can be customized to provide additional services similar to those provided by PBXs.

Telephones available today are modular; that is, they are equipped with a connector plug that fits into a standard jack or outlet, enabling the customer to attach them to a building's internal wiring. Thus, they can be unplugged and taken to a service center for testing in case of trouble. Rotary-dial telephones are being replaced by telephones equipped with pushbuttons, which are easier to use and speed up the calling process.

Other types of telephones that are gaining in popularity are cordless, remote radio-controlled telephones, which allow a user to conduct a telephone conversation a short distance from the regular telephone instrument; speakerphones, which allow the user to listen and talk without a handset; and mobile telephones, which provide telephone service between stationary telephones and a moving vehicle.

Review Questions

1. Describe how the telephone works.
2. What is the function of the central office?
3. What is the function of a modem? Why is it necessary to use a modem?
4. Discuss the nationwide numbering plan. How many telephone numbers can there be in a given geographical area?
5. What is the purpose of the world numbering plan?
6. What is a private branch exchange?
7. Discuss key telephone sets, including their main functions and principal service features.
8. Describe some of the service features available on computerized PABXs.

References and Bibliography

Edwards, Morris. "Local Loop Bypass Paves Way for Wideband Services." *Communications News*, September 1983, 50–54.

Martin, James T. *Telecommunications and the Computer*, 2d ed. Englewood Cliffs, N.J.: Prentice-Hall, Inc., 1977.

Mueser, Roland, ed. *Bell Laboratories Innovations in Telecommunications*. Murray Hill, N.J.: Bell Laboratories, Incorporated, 1979.

Schloss, Jason S. "Analog vs. Digital PBXs: Weighing Their Pluses and Minuses." *Telephony*, May 2, 1983, 84–89.

Technical Staff, Bell Telephone Laboratories, Incorporated. *Engineering and Operations in the Bell System*. Murray Hill, N.J.: Bell Telephone Laboratories, Incorporated, 1977.

Telecommunication Networks

5

The concept of networks is familiar to all of us. We use highway networks of country, intrastate, and interstate roads to take us from one location to another. Trains and airplanes greatly enhance their usefulness by being linked into a network. Ships sail in a network of navigable rivers and canals. The U.S. Postal Service operates a network to deliver mail to every state, city, and home in the nation and to link up with postal systems of other nations. Radio and television networks such as the American Broadcasting Company (ABC), National Broadcasting Company (NBC), and Columbia Broadcasting System (CBS) use facilities of numerous local broadcasting stations.

Brenda Maddox, writing in *Beyond Babel*, describes the world's telephone networks as "technical miracles." She says, "The networks can make millions out of billions of possible connections among telephones thousands of miles apart in a matter of seconds. And they can just as swiftly unmake them and make millions of new paths in quite different directions."[1]

Similarly, a telecommunication network interconnects a number of stations using telecommunication facilities (Figure 5.1). The network consists of transmission systems, switching systems, and station equipment. The physical circuit between two points is referred to as a *link*. A *node* is a point of junction of the links in a network. A basic principle of networking is that each of the stations in the network must be able to connect with any of the other stations in the network. The stations on a network intended for voice communications are, of course, telephones. Stations on data networks may be telephones or other devices such as teletypewriters, facsimile units, or computers.

1. Brenda Maddox, *Beyond Babel* (New York: Simon & Schuster, 1972), p. 202.

Figure 5.1
A Simple Local Network

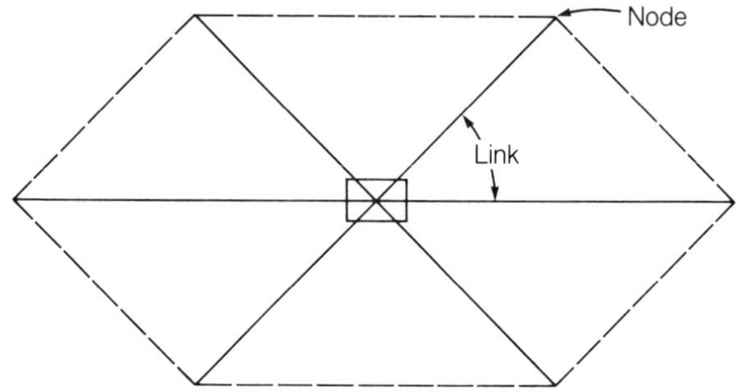

This chapter will deal primarily with long distance networks, since these intercity networks embody all of the fundamental principles of networking.

Public and Private Telecommunication Networks

Telecommunication networks may be public or private. They may be designed for voice, data, message, or image transmission. They may be local or global—intercity, interstate, or intercontinent—in scope. They may use any of a number of transmission channels: open wire, paired cable, coaxial cable, radio, satellite, waveguide, or optical fibers. Regardless of their differences, the electronic principles that make these networks work are common to all types of telecommunication networks.

There are many telecommunication networks in operation, and they provide a wide variety of services. Some of these are public-switched networks (PSN), private-line voice networks, audio program networks, video program networks, private-line data networks, packet-switching networks, and public-switched data networks.

Most networks fall into two main categories: the public-switched networks and private- or leased-line networks. Unfortunately, each of these two classifications is also known by a number of other names, a fact that causes some confusion. For example, the following network terms are synonymous: *public switched, public, switched, message toll service (MTS), long distance, direct-distance dialing (DDD),* and *inter-exchange facilities.* Similarly, the terms *private line, leased line, dedicated line, full-time circuit,* and *tie line* are used interchangeably.

Public-Switched Networks

Public-switched networks provide business and residential telephone service for voice and data transmission to the general public. Users share common switching equipment and channels; thus, callers wait their turn for service if all the facilities are in use. Fees paid for the use of the network are assessed on a per-call, per-minute, per-mile basis. Public-switched networks are by far the largest category of networks in terms of volume of traffic and revenues.

Until the late 1960s the Bell System was the only communications common carrier offering nationwide voice network services to the public. When a long distance call was initiated by an independent telephone company subscriber, it was switched to Bell System facilities for completion.

The MCI decision opened the way for other communications common carriers to establish point-to-point communication routes and combine them into networks. The decision also ordered the Bell System to provide interconnect privileges to the specialized common carriers, thereby allowing them to compete directly with the Bell System for long distance business.

Initially, SCCs offered service only for private or leased-line facilities. Later they expanded to include public-switched long distance services. Today, the public can choose from a number of SCCs such as MCI, RCA, Western Union, and SPC, which offer private- or public-switched network services.

Packet-Switching Networks Public packet-switching networks were originally designed to provide a more efficient method of transferring data over networks. They are still used primarily for this purpose; however, digitized voice may also be transmitted using packet-switching techniques.

In packet transmission a data message is divided into discrete units called packets that are routed individually over the network. Since each of the various packets of the message can be routed over different transmission facilities, they may arrive out of sequence. Each packet contains control information that enables the message to be reassembled in proper sequence before it reaches its final destination. Packet switching is efficient because packets use the network only for the brief time they are in transit, in contrast to a circuit-switched message that requires the use of the line for the duration of the message. The concept of packet switching is analagous to computer time sharing; in each case several users share the same facility in what appears to be the same time.

Packet-switching networks are described as value-added networks because the transmission lines are supplemented with computerized switches that control traffic routing and flow. A standard feature of packet switching is automatic error detection and correction of transmitted packets.

The first packet-switched network was ARPANET, the U.S. Defense Department's Advanced Research Projects Agency Network,

Figure 5.2
Foreign Exchange Service

Calls can be made at local call rates to and from another exchange

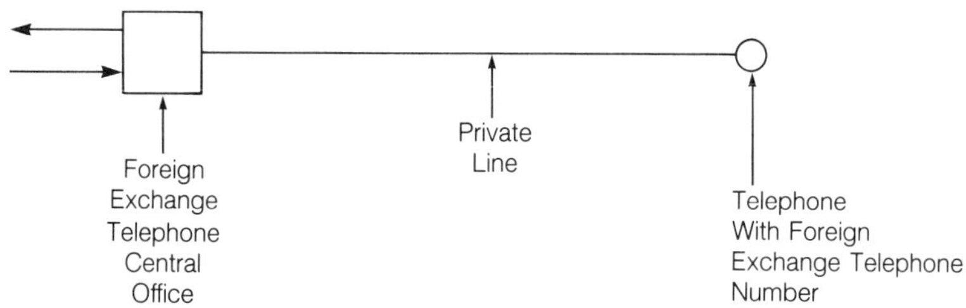

which is used primarily by government and academic institutions. Several vendors currently offer packet-switching services for data communications, including Telenet, Graphnet, Tymnet, ITT, and AT&T Information Systems.

Private Networks

Private or leased facilities are dedicated to the exclusive use of one subscriber; the line—or network of lines—is always available for the customer's use. The subscriber pays a flat monthly fee for the service and is entitled to use the lines on an unlimited basis.

Foreign Exchange (FX) Service (Figure 5.2) is service in a telephone exchange that is "foreign" to (outside) the one in which the user is located. This service is implemented by a private leased line that connects the subscriber's station (telephone, PBX, or modem) to a central office in another (foreign) exchange. The subscriber can then make an unlimited number of calls to any number associated with the foreign exchange area for just the cost of a local telephone call. The subscriber is given a directory listing in the foreign area directory. FX is a two-way service, so people from the foreign exchange area may also call the subscriber using what to them is a local telephone number. The call goes over the subscriber's FX line; the caller does not pay long distance charges.

Subscribers who have substantial call volume to a foreign exchange location may find it advantageous to obtain FX service. For example, a firm located in Lansing, Michigan, might use FX lines to Detroit if much of its business were there. Some large city department stores have FX lines to suburban locations to encourage business from suburbanites by permitting them to call the main store toll free.

Permanently connected private lines have several advantages. For one thing, the line—or network of lines—is always available; there are no busy signals. Another advantage is that private lines can be *conditioned* or specially treated to reduce distortion and improve transmission quality. Lines are conditioned by the addition of electronic components to the circuit. The customer can select the grade of transmission quality, as defined by the carrier's specifications, that best meets transmission requirements. Conditioning results in fewer errors during transmission, fewer echos, and less *crosstalk*, a condition in which one pair of wires picks up the transmission on adjacent wires and a conversation becomes faintly audible. Further, lines that have been conditioned are capable of higher transmission speed, thereby sending or receiving more information in a given amount of time. Line conditioning is a service offered by long distance carriers; it can be performed only on private voice-grade lines.

The cost effectiveness of leased lines depends upon the amount of time subscribers use the lines and the mileage covered by the circuit. If the lines will be used only briefly each day, a public-switched network will probably be more economical since leased lines are charged at a flat rate regardless of time usage. However, subscribers requiring considerable transmission time often realize substantial savings by using leased lines.

Because of their many advantages, large companies and government organizations frequently use leased lines. Private facilities range from point-to-point telephone lines to nationwide switched voice and data systems such as Common Control Switching Arrangement (CCSA). CCSA is a private long distance network that offers intercity circuits under a special pricing arrangement. The network trunks and access lines are private lines; however, the arrangement uses switching equipment at company exchanges to switch calls in the private network. Stations connected to the network may call one another without using the public toll facilities. One of the largest private leased networks is the Automatic Voice Network (AUTOVON), serving the U.S. Department of Defense. The network spans the United States and extends to overseas and Canadian locations carrying data as well as voice.

Local Area Networks The term *local area network (LAN)* describes a configuration of telecommunications facilities designed to provide internal communications within a limited geographical area. The network can be used to interconnect telephones, computers, terminals, word processors, facsimile machines, and other office equipment within a building, a building complex, or a metropolitan area. LANs usually make use of the latest technologies to meet the needs of a data communications-intensive environment. They can also be designed to handle combinations of data, voice, and video transmission.

Transmission Media

Transmission systems are the links that interconnect the nodes of a telecommunication network. They may consist of any one of a variety of transmission media. For many years the principal media used for telecommunications transmission were wire conductors: open-wire pairs, wire-pair cables, and coaxial cables. Today, a number of other media such as microwave radio, satellites, waveguides, and optical fibers are in use.

Open-Wire Pairs

Early telephone and telegraph transmission lines consisted of open-wire pairs. The lines contained uninsulated wire suspended between poles with cross arms. The wires were either copper, copper-clad steel; or galvanized steel, and they required about 12 inches of separation to prevent momentary shorts.

Open-wire pairs served the nation's telecommunication needs for many years. However, they had several inherent disadvantages:

1. They were very susceptible to storm damage; thus, their maintenance cost was high.
2. Open-wire pairs were also subject to crosstalk.
3. Open-wire pairs were unsightly; as more and more telephones were installed, the number of open-wire circuits strung on poles in the large cities reached the saturation point (Figure 5.3).

Figure 5.3
A City Street with Open-Wire Pairs

(Reproduced with permission of AT&T)

Figure 5.4
Open-Pair Wires

Today, it would be impossible to meet telephone requirements with open-wire facilities. (See Figure 5.4.)

Wire-Pair Cables

Wire-pair cable consists of copper conductors insulated by either wood pulp or plastic and twisted into pairs (Figure 5.5). Twisting minimizes the interference between pairs when they are packed into a large cable. Each of the two-wire circuits is capable of carrying one telephone channel. The cables may be either suspended from poles or buried underground. Wire-pair cables have largely replaced open-wire pairs.

Coaxial Cables

A coaxial cable consists of one or more hollow copper cylinders with a single wire conductor running down the center. A single cable can carry a very large number of telephone calls. The name "coaxial" is derived from the fact that the cylinder and the center wire each have the same center axis (Figure 5.6). Because coaxial cables transmit at

Figure 5.5
A Twisted Wire-Pair Cable

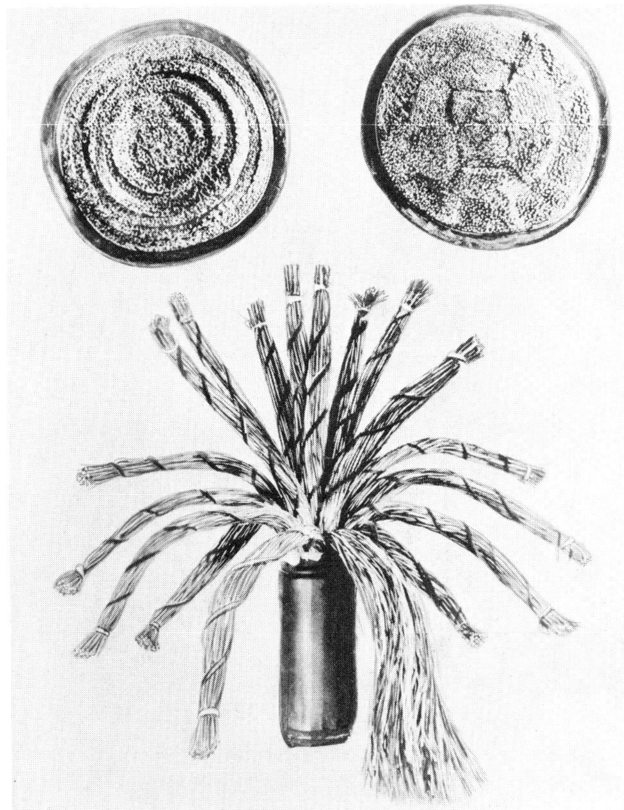

(Courtesy of The Western Electric)

Figure 5.6
A Coaxial Cable

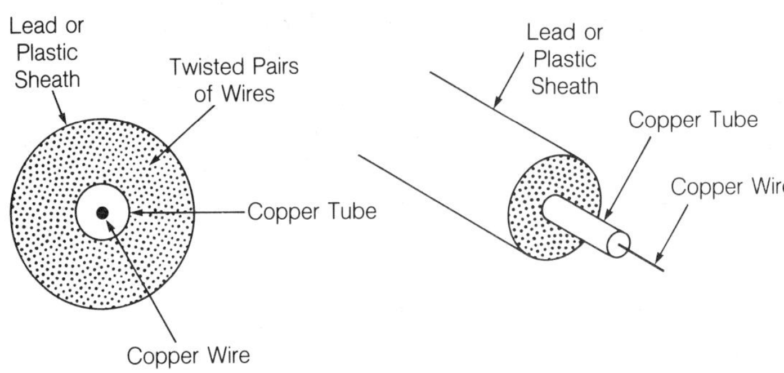

very high frequencies, they have little distortion or signal loss. Therefore, they are a better transmission medium than either open-wire pairs or wire-pair cables.

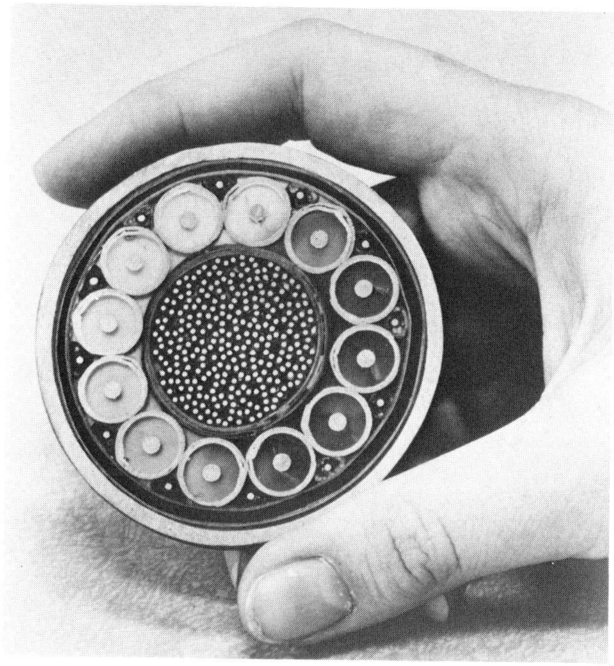

Figure 5.7
A 12-Unit Coaxial Cable

(Reproduced with permission of AT&T)

Figure 5.7 shows a cable composed of 12 coaxial units surrounding a core of conventional wire conductors. It can handle approximately 1,000 telephone conversations simultaneously.

Figure 5.8 shows a sample length of Co-ax 20, a 20-unit coaxial cable capable of carrying 18,740 telephone calls at once.

Microwave Radio

A microwave radio system sends signals through the atmosphere between towers usually spaced about 20 to 30 miles apart, amplifies them, and retransmits them at each receiving station until they reach their destination. Microwave radio operates at the high-frequency end of the radio spectrum. The signals follow a line-of-sight (straight-line) path, and the relaying antennas must be within sight of one another. (See Figure 5.9.) One of the principal problems affecting microwave radio is the variation in signals caused by changes in atmospheric conditions. Moisture and temperature conditions can cause the radio beam to bend, resulting in fading.

The principal advantage of microwave radio is that it can carry thousands of voice channels without physically connected cables between points of communication, thus avoiding the need for continuous right-of-way between points. Further, radio is better able to

Figure 5.8
A 20-Unit Coaxial Cable

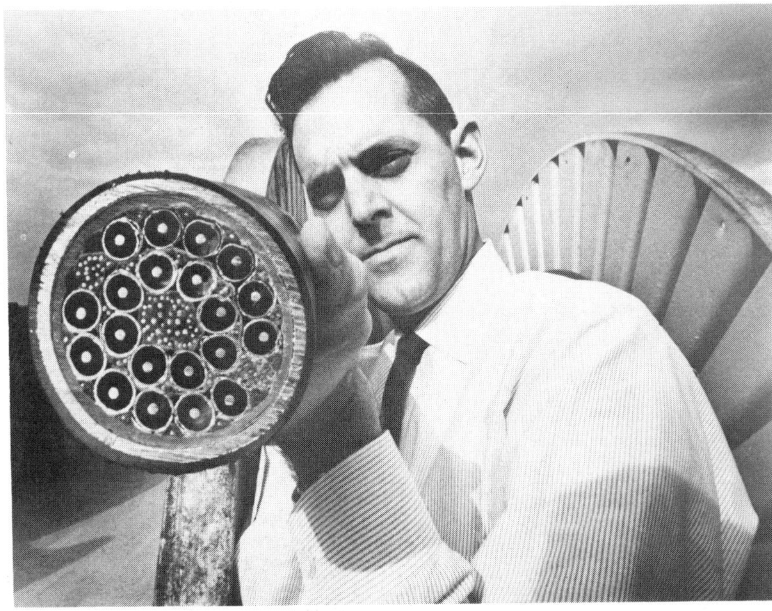

(Reproduced with permission of AT&T)

Figure 5.9
A Microwave Installation. A microwave relay tower located in the desert. All microwave antennas are on towers within sight of one another. Relay towers are usually spaced 20 to 30 miles apart.

(Courtesy of MCI Telecommunications Corporation)

(Courtesy of MCI Telecommunications Corporation)

Figure 5.10
A Microwave Installation. A microwave relay tower located on a mountain range. Since microwave transmission is line-of-sight, the towers are placed on elevated points when possible.

span water, mountains, or heavily wooded terrain that pose barriers to wire or cable installation. Most long distance links today are either coaxial cable or microwave radio. (See Figures 5.10 and 5.11.)

Communication Satellites

Communication satellites relay microwave transmissions. The satellite functions as a microwave tower located high in the sky. The *transponder* is equipment that receives a signal, changes its frequency so that the outgoing signal does not interfere with the incoming signal, and retransmits it. Satellites with solar-powered batteries orbit directly over the equator at a distance of 22,300 miles above the earth so that they travel at exactly the same speed as the rotation of the earth. Because of their very high altitudes, they can receive radio beams from any location in the country and reflect them back to a large portion of the world (Figure 5.12). Their very high altitude also permits them to overcome obstacles that block line-of-sight transmission (microwave radio) such as mountains and the curvature of the earth. Satellites can handle very large volumes of voice and data transmission simultaneously.

Figure 5.11
A Microwave Installation. Satellite antennas at Western Union's Glenwood, New Jersey, earth station capture signals from satellites. The microwave tower is used to send messages to other microwave towers.

(Courtesy of Western Union)

Figure 5.12
Satellite Coverage of Earth's Surface

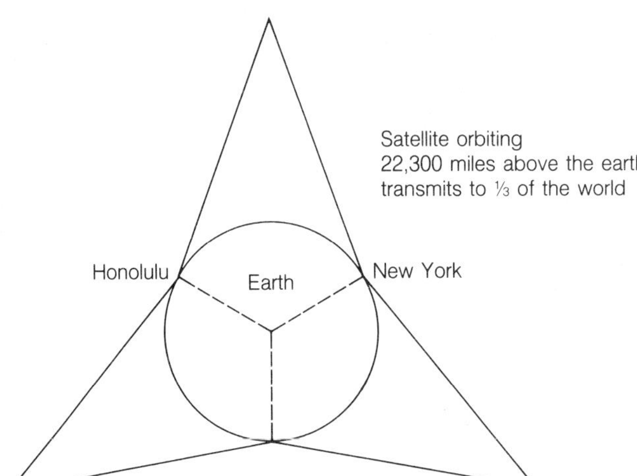

An efficient antenna system is a vital part of any radio system—microwave or satellite. The transmitting antenna radiates as much of the transmitter's energy as possible toward the receiver, and the receiving antenna collects as much of this energy as it can. A satellite antenna, or *earth station,* consists of a large dish that points at the satellite in much the same way that a microwave relay tower points to the next tower in the relay chain. Probably most of us have noticed the satellite dishes that are appearing atop buildings or on the ground in ever-increasing numbers.

Satellite communication has several unique characteristics that make it very attractive from both performance and economic standpoints. These are:

1. The cost of a satellite is not dependent upon the distance between stations.
2. Natural barriers such as mountains, oceans, or densely wooded terrain do not impede placement of satellite facilities.
3. Signals are available at any point within the satellite's path.

Where traffic moves between a few nodes of a network over great distances, satellite communication is the most cost-effective medium because the cost of the link is not dependent upon distance. In fact, the greater the distance, the greater the cost advantage.

Natural barriers make it difficult, and therefore expensive, to construct any terrestrial or submarine transmission system. Satellite systems effect substantial cost savings by avoiding these barriers.

Since a signal is available at any point within the satellite's path, the only additional cost for a new receiving node is the cost of the hardware. Thus, in the case of point-to-multipoint distribution, the satellite has a significant cost advantage. Further, the addition of new receiving nodes greatly increases the cost effectiveness of the system.

A number of businesses provide voice and/or data transmission service by satellite (Figures 5.13 and 5.14). These companies either have their own satellites or purchase capacity from another organization. Some companies offering satellite transmission services are RCA Americom, Satellite Business Systems (SBS), Western Union, American Satellite Company, GTE Satellite Communication System, Hughes Communications, Inc., and Communication Satellite Corporation.

The distribution of printed text has been changed considerably by the use of satellites. *The New York Times, The Wall Street Journal, USA Today,* and other national newspapers use satellite communications to reduce their distribution costs. The copy is transmitted in facsimile form to a number of printing plants in various regional locations.

Figure 5.13
WESTAR IV Satellite

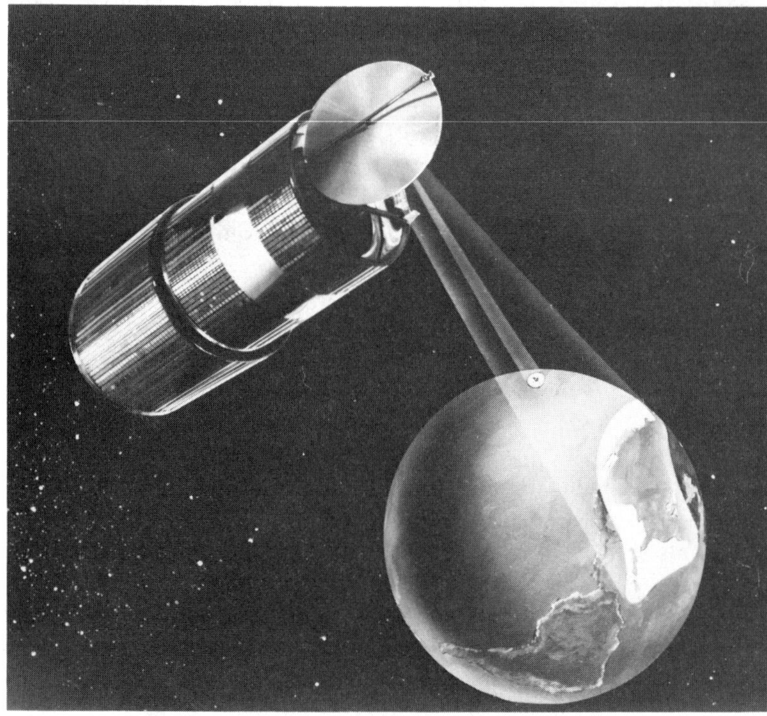

(Courtesy of Western Union)

Waveguides

A *waveguide* (Figure 5.15) is a rectangular or circular metal tube down which very high frequency radio waves travel. The metal tube confines the radio waves and channels them to a point where they are released into the air to continue their travel over microwave transmission facilities. Waveguides can transmit greater amounts of power with less energy loss than coaxial cables.

Rectangular waveguides have been used for some time to connect microwave transmitting equipment to the microwave towers. Their use is generally limited to distances of less than a thousand feet. The newer and more efficient waveguide is the circular type, which consists of a precision-made pipe about two inches in diameter. This type of waveguide is capable of transmitting frequencies much higher than rectangular waveguides.

The waveguide is attractive because of its wide bandwidth and low-loss transmission characteristics. Deterrents to its use are its critical engineering requirements and very high cost.

Waveguides have great potential for further use in long distance communications. However, the development of optical fibers has resulted in a similar, but somewhat superior, medium at appreciably lower costs.

(Courtesy of Bell Laboratories)

Figure 5.14
Telstar I Satellite

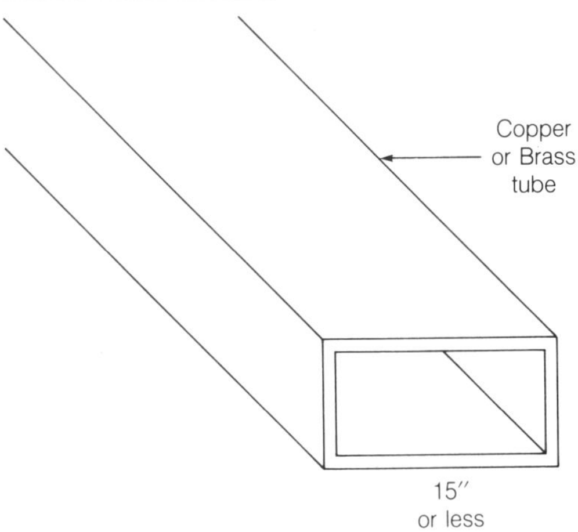

Figure 5.15
Waveguide

Optical Fiber Systems

The newest and most remarkable technological development in transmission media has been fiber optic waveguides. *Fiber optics* are hair-thin filaments of transparent glass or plastic that use light

Figure 5.16
Cross-Section of Lightguide Cable

(Courtesy of Western Union)

instead of electricity to transmit voice, video, or data signals. The fibers act as waveguides and have the potential of carrying an extremely high bandwidth with low signal attenuation. Covered to prevent light loss along the line, the fibers are bundled together into a flexible cable (see Figure 5.16). In most applications, optical fiber systems combine optical fibers with *laser* (an acronym for light amplification by stimulated emission of radiation) technology to generate very high frequency beams of light with tremendous information capacity. Some optical fiber systems use light-emitting diodes (LEDs) as a source of light.

In lightwave communications, the transmission sequence begins with an electrical signal. The signal is transformed into a light signal by a laser or other source that couples the light into a glass fiber (Figure 5.17) for transmission. The light signal is renewed along the way by a repeater unit. At its destination the light is sensed by a receiver and converted back into electricity.

Optical fiber systems have many advantages:

1. The hair-thin size of the fiber lightguides makes possible cables that are much smaller and lighter than their wire counterparts (Figure 5.18).
2. Lightwave systems cost less to manufacture than wire systems of equivalent capacity.

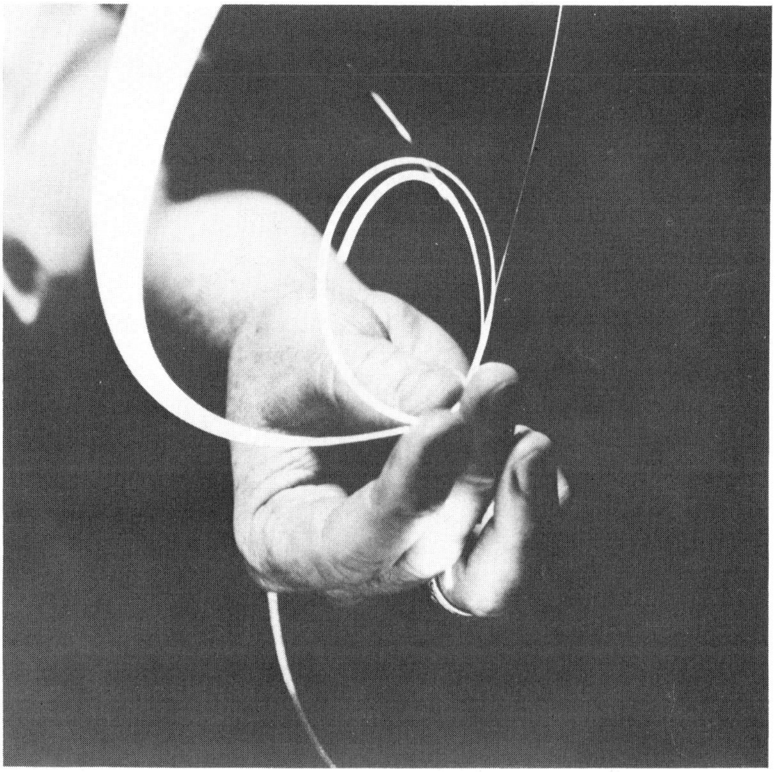

Figure 5.17
Glass Fibers

(Courtesy of Bell Laboratories)

3. Transmission through lightguides is relatively immune to the effects of lightning and other types of electrical interference.

One of the first commercial applications of this technology was a digital lightwave system called FT 3, which transmits 44.7 million bits of information per second. It began service in September 1980 and covered a 6.5-mile route of the Bell System linking three central offices in Atlanta, Georgia. AT&T, MCI, and other carriers are actively expanding their networks by the construction of fiber-optic facilities. See Figure 5.19.

Transmission Channels

Information travels from one point to another along a transmission link that carries an electrical signal. The link is called by any of a number of different names: *channel, trunk, path, circuit,* or *facility.*

In designing a voice or data network, it is necessary to decide which transmission medium will be used. Any of the transmission

Figure 5.18
Cables of Glass Fibers

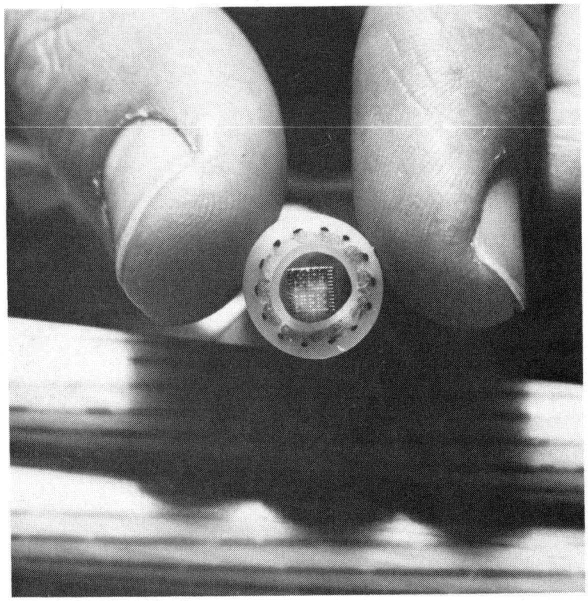

(Courtesy of Bell Laboratories)

Figure 5.19
Lightguide Installation on the Golden Gate Bridge

(Courtesy of The Western Electric)

Figure 5.20
Intercity Telephone Call Media

Route	Medium
Subscriber telephone to local central office	Wire-pair cable
Local central office to toll switching office	Wire-pair cable, coaxial cable, or fiber optic
Toll switching office to distant city toll office	Coaxial cable, microwave radio, satellite, or fiber optic
Distant city toll office to local central office	Wire-pair cable, coaxial cable, or fiber optic
Local central office to called party's telephone	Wire-pair cable

media discussed in the previous section can be used to move information. Also, two or more types of transmission media can be combined into a network.

Practically every long distance telephone call travels over a variety of transmission media before it reaches its destination. Figure 5.20 illustrates the media that could be used to route an intercity telephone call.

Another important choice to make is the type and grade of circuit to be used. Here it is necessary to determine whether communication lines must transmit in one direction only or in both directions. If transmission will occur in both directions, the system's designer determines whether the transmission in both directions will be simultaneous. The principal criteria used in choosing the media and equipment are transmission requirements and economics.

Attenuation and Repeaters

As a signal travels along a transmission line, there is a natural loss of power, that is, the signal grows weaker. This loss in power is called *attenuation*. If the signal is not strengthened somehow, it will not reach its destination in a way that it can be understood. To compensate for this loss of power, amplification devices known as *repeaters* are inserted into the channel. A repeater is a device used to restore signals that have been distorted because of attenuation to their original shape and transmission level. Repeaters are also known as *amplifiers* since they amplify the signal while it still has enough magnitude to represent the original signal. Repeaters placed at equal distances throughout the transmission channel reinforce the signal strength so that when the signal is received at the terminating point it can be fully understood. Attenuation increases as frequency of the transmission bandwidth increases; it also increases as the diameter of the wire used in the circuit decreases.

Echo and Its Control

To provide high-quality speech transmission over long toll connections, a type of distortion known as *echo* becomes important. Echo is an electric wave that has been reflected back to the transmitter with sufficient magnitude and delay as to be perceived. Echo is a function of distance on the circuit; the greater the distance, the more time it takes for the echo to return to the person talking.

Echo is caused by a change in impedance on the transmission line, causing the signal to be reflected at reduced amplitude. The resulting "hollow" sound can be objectionable; therefore, the telephone company installs *echo suppressors* to reduce it to a negligible level. Echo suppressers are devices that detect speech signals transmitted in either direction on a four-wire circuit. The suppressors introduce loss in the opposite direction of the speech transmission to suppress echos. In the public telephone network, echo suppressors are used typically in trunks longer than 1,850 miles.

Echo suppressors designed for voice cannot be used when data is transmitted over voice circuits, since the speech detector detects only speech, not other sounds. To transmit data over voice circuits, the echo suppressors must be disabled. This is accomplished by installing *echo suppressor disablers,* devices that transmit a tone that can be heard on the telephone as a high-pitched whistle. The tone puts the echo suppressor out of action until there has been no signal on the line for a period of approximately 50 milliseconds.

Echo suppressors have two main drawbacks:

1. They tend to clip the speech as they open and close the transmission path.
2. They have no capability for suppressing echo during two-way transmission, because both transmission directions cannot be opened simultaneously.

The newest technique to control echo is a device known as an *echo canceler.* Because these devices eliminate both of the above-cited problems, they are widely used in modern circuit construction.

Modes of Transmission

There are three basic modes of transmission: *simplex, half duplex (HDX),* and *full duplex (FDX)* (Figure 5.21).

Simplex This mode of operation is used when transmission is in one direction only. Simplex circuits are *two-wire* circuits; however, transmission is in the same direction at all times. An analogy would be a radio announcer who talks to listeners who cannot reply.

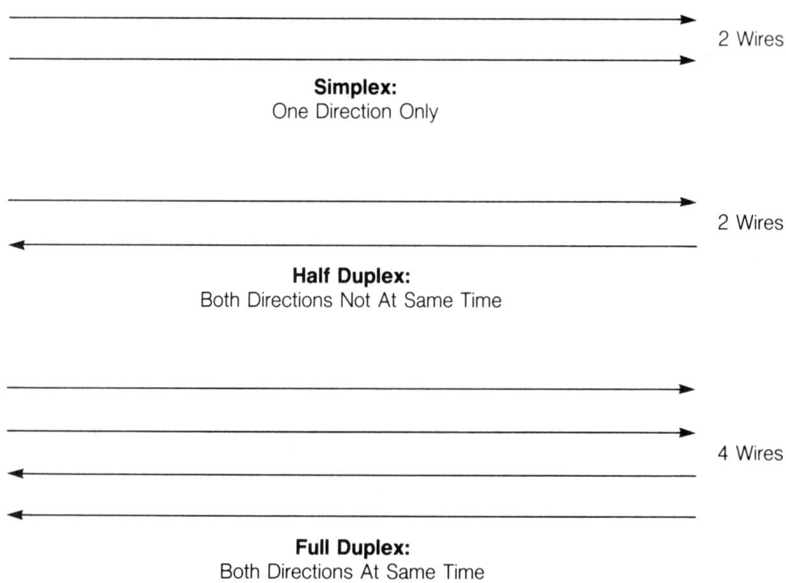

Figure 5.21
Simplex, Half-duplex, and Full-duplex Circuits

HDX In this mode of operation, transmission can be in either direction, but in only one direction at a time. HDX is often referred to as two-wire transmission. One station transmits information and only when that transmission has been completed can the other station respond. This could be compared to two debaters; each may speak, but only in turn and one at a time.

FDX This mode of operation is used for transmission in both directions simultaneously. Full-duplex circuits are sometimes referred to simply as duplex circuits. There is no way that human communications can be compared with full-duplex transmission because when two people interrupt each other by talking at the same time, intelligent information cannot be conveyed.

Historically, the communications circuit had to be four-wire for full-duplex capability. However, two-wire circuits can be used in full-duplex mode if the amplifiers are designed to permit transmission in both directions simultaneously or if a modem is used to split the band of frequencies into two parts.

Uses of Each Transmission Mode Simplex circuits have limited utility. Their principal use is in signaling, where they are used to control the telephone switching process. They are not used in voice or data transmission. Either half-duplex or full-duplex circuits may be used for data transmission. With leased telephone lines the user can choose between half-duplex and full-duplex circuits. Full-duplex circuits cost more than half-duplex circuits; however, when data communications require two-way simultaneous transmission, a full-duplex circuit is necessary.

Alternating Current

There are two types of electrical current: *direct current (DC)* and *alternating current (AC)*. Direct current travels in only one direction in a circuit (+ to −), while alternating current travels first in one direction (+ to −) and then in the other direction (− to +). Household current is alternating current—it reverses itself continuously. It is known as 60-cycle electric current because it reaches a maximum flow in one direction, then reverses itself to a maximum flow in the opposite direction; the reversal repeats back and forth, making 60 complete cycles in one second.

The number of cycles completed per second is the *frequency* of the current. It is expressed in Hertz (Hz). The higher the frequency, the more direction reversals each second. *Amplitude* is the amount of variation of an alternating quantity from its zero value. It is sometimes described as volume, intensity, or loudness. Frequency and amplitude are the two parameters of any electronic signal.

The frequency of oscillation of an alternating current can be represented as an undulating wave known to mathematicians as a *sine wave*. A sine wave can be drawn by plotting the power level of the alternating current against intervals of time (Figure 5.22). The alternating current starts out at zero power level, rises to a maximum at ¼ of a second, then drops back to zero after ½ of a second and continues to a maximum in the other direction at ¾ of a second and back to zero power at one full second.

Bandwidth

Sounds that we hear consist of a mixture of electromagnetic waves that usually range from about 300 Hz to 3,300 Hz, although the range extends up to 20,000 Hz for the person who has exceptional hearing ability. However, since most voice characteristics are contained in sound waves in the range from 300 to 3,300 Hz, telephone lines are designed to transmit frequencies in this range. *Bandwidth* means the range of frequencies that are transmitted (Figure 5.23). It is the difference in Hz between the highest and lowest frequencies of a band. Thus, the *voiceband* is said to be 3,000 Hz, or the difference between 300 Hz and 3,300 Hz.

Figure 5.22
Sine Waves of Differing Periods

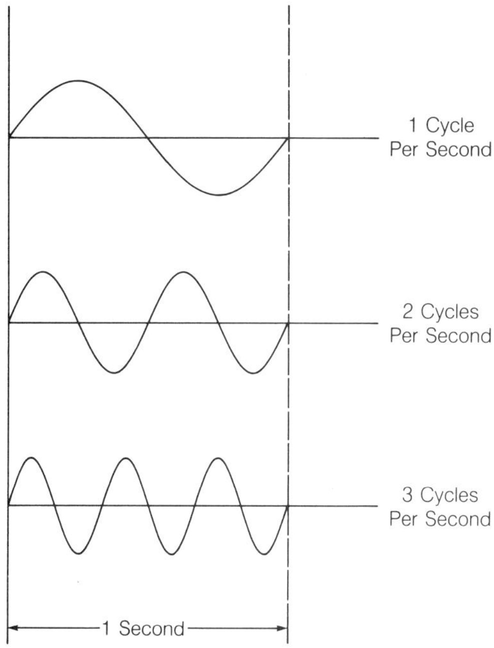

Figure 5.23
Bandwidth of Voice Transmission Over Telephone Lines

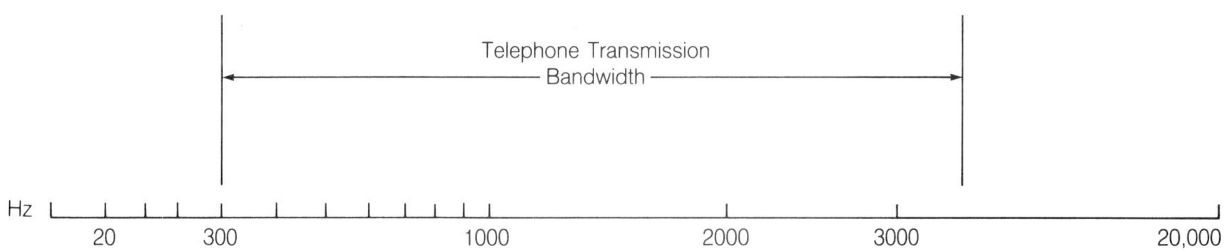

The chief difference between the grades of available transmission channels is their bandwidth. In classifying channels by bandwidth, there are three general categories:

1. *narrowband* (0 to 300 Hz)
2. *voiceband* (300 to 3,300 Hz)
3. *broadband* or *wideband* (over 3,300 Hz)

Narrowband channels are used for nonvoice service such as teletypewriter and low-speed data transmission. Voiceband channels are used for voice transmission, foreign exchange (FX) service, and data communications. Voiceband lines are the most prevalent form of communication facilities. Wideband channels are used for high-speed data, facsimile, and video transmission. Their principal value, however, is for multiplexing many channels of various types onto a single bandwidth.

Frequency

The *frequency* of an oscillating current—the number of complete oscillations the wave makes per second—can be controlled so that the current oscillates at a specific frequency. This frequency produces a given sine wave, which carries signals representing voice or other forms of communication.

Although telecommunication transmission systems are quite different from commercial radio systems, they both work on many of the same principles. A radio station broadcasts (transmits) at an assigned frequency. A listener can tune the radio to receive a particular station's transmission; the radio tuner selects the desired station by locating the frequency on which the station is operating. Each station is received on a different frequency. However, the sound is in the same bandwidth for all stations (300–3,300 Hz).

Similarly, a telecommunication channel is designed to operate at a particular frequency. This frequency serves as a carrier that transports specific communication bandwidths. Telecommunication systems differ from radio systems in that the signals transmitted are confined to the medium (wire, cable, coaxial cable, microwave, satellite, or fiber optic) rather than broadcast to the air.

Modulation and Demodulation

Modulation is the process of changing the form of a signal carrying intelligence (voice or data) to make it compatible with a different transmission medium. Modulation alters (modulates) a signal to enable it to be transmitted on a different frequency. Modulation improves the efficiency of a transmission by increasing the transmission capacity of a telecommunications channel through the use of higher frequencies.

The newer transmission media, particularly satellites and fiber optics, have wider bandwidths than earlier transmission media. In addition, newer and more efficient transmission techniques, such as pulse code modulation (PCM), have been developed. These factors have significantly increased the number of messages that can be carried by each channel.

In the transmission process, modulation is a function of imposing the signal that carries the intelligence of a message onto the carrier

Figure 5.24
Amplitude Modulation

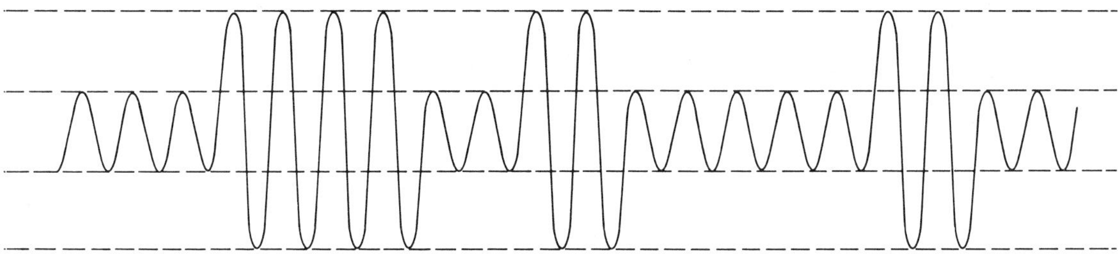

Figure 5.25
Frequency Modulation

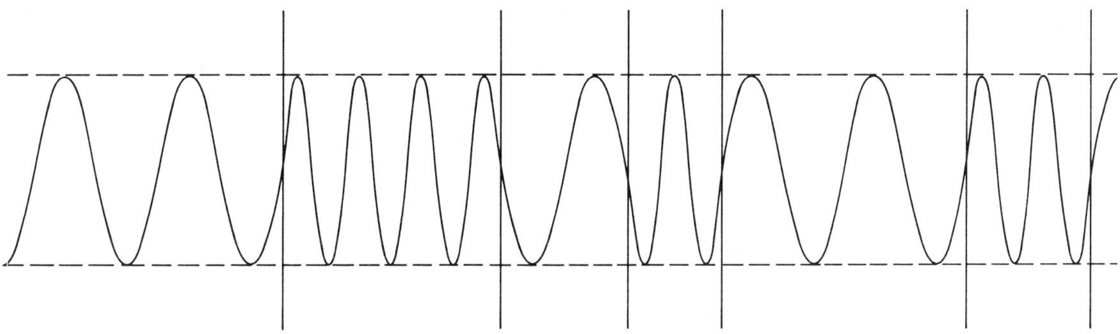

wave generated by the flow of alternating current. *Demodulation* changes the bandwidth back into the form of the original message signal.

There are two principal types of modulation: amplitude modulation and frequency modulation. *Amplitude modulation* (Figure 5.24) is the process whereby the intelligence of the signal is represented by variations in its amplitude or strength. *Frequency modulation* (Figure 5.25) is the process whereby the intelligence of the signal is represented by the variations in the frequency of the oscillation of the signal.

There are many other forms of modulation. Two forms that are used frequently in data transmission are phase modulation and pulse code modulation.

Phase modulation is used in equipment operating at high speeds; it makes use of a frequency phase shift that occurs when a specific pattern of digital information is received. *Pulse code modulation* is one of the newest forms of transmission. This system uses a sampling process and requires specific types of modulation/demodulation equipment as well as special transmission facilities. These two types of modulation will be described more fully in Chapter 7.

Figure 5.26
Multiplexing

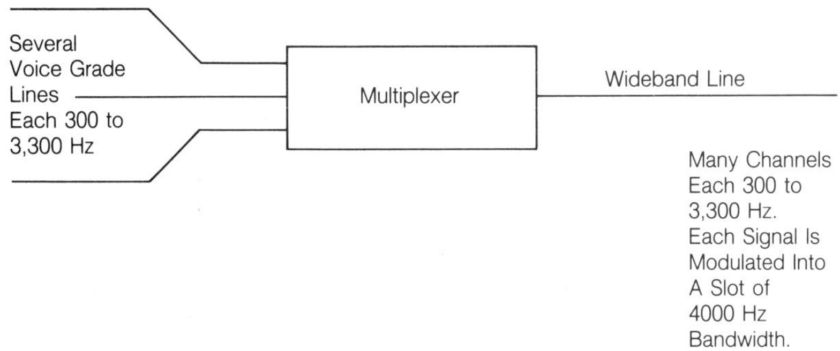

Multiplexing

Multiplexing is the division of one transmission channel into two or more individual channels. It is accomplished by a *multiplexer* (Figure 5.26), a device that combines a number of low-speed channels into one higher speed channel at one end of a transmission and divides them into low-speed channels at the other. Multiplexers must be provided at both ends of the circuit. The process permits a voice-grade channel to carry two or more narrowband channels. Similarly, a wideband channel can carry several voiceband channels. Multiplexing reduces line costs by increasing the number of lines that can be carried by a communications channel.

There are two basic multiplexing techniques: frequency-division multiplexing (FDM) and time-division multiplexing (TDM). *Frequency-division multiplexing* divides the channel frequency range into narrower frequency bands. *Time-division multiplexing* assigns a given channel successively to several different users at different times.

Summary

Interconnecting a number of individual telephone lines into an integrated network makes it possible for any telephone to be connected to any other telephone anywhere in the world. Services such as distributed data processing and electronic mail depend on the existence of networks.

A telecommunication network consists of transmission systems, switching equipment, and station equipment. A basic principle of networking is that each of the stations in the network must have the capability of establishing a connection with any of the other stations in the network.

The stations on a voice network are telephone sets. Stations on a data network may be telephones, teletypewriters, facsimile machines, or computers. Transmission channels may be open wire, paired cable, coaxial cable, radio, satellite, waveguide, or optical fiber. The newer transmission media, such as satellites, waveguides, and optical fibers, make it practical to offer service to locations that would otherwise be inaccessible.

The two types of telecommunication networks are public and private. Public-switched networks provide telephone service to the general public. They are by far the largest category of networks in terms of volume and revenue.

Public packet-switching networks provide an efficient method of transferring data. In this technique, messages are divided into packets for transmission. The transmission channel is only occupied during the transmission of the packet; afterward it is available for the transmission of other packets. This contrasts with circuit switching, wherein entire messages are transmitted as a unit, and the transmission channel is occupied for the duration of the transmission.

Private or leased networks are dedicated to the exclusive use of one subscriber. Both public and private networks are provided by AT&T, GTE, MCI, RCA, Western Union, and many other common carriers. There are three basic modes of transmission: simplex, half-duplex, and full-duplex. Voice transmission requires full-duplex circuits. The chief difference between the grades of available transmission channels is bandwidth. The three categories of bandwidth are:

1. narrowband, used for teletypewriter and low-speed data
2. voiceband, used for voice and data transmission
3. wideband, used for high-speed data, facsimile, and video transmission

The two characteristics of electrical current are amplitude and frequency. Amplitude measures the strength of the signal; frequency is the number of times a wave form repeats itself in a given period of time.

Modulation is the conversion of signals from pulse form to wave form. Demodulation is the reversal of the modulation process.

Multiplexing is the division of a communication channel into two or more narrower channels. This process is performed by an electronic device known as a multiplexer. Two multiplexers are required—one at each end of the circuit.

Review Questions

1. What three elements comprise a telecommunication network?
2. What is the basic principle of networking?
3. Briefly describe packet switching. Why is it an extremely efficient method of transferring data?

4. Under what condition would it be advantageous for a subscriber to obtain FX services?
5. Cite two advantages of private telephone lines.
6. What are some new types of transmission media? Briefly describe their advantages.
7. What transmission medium would you use if you were planning a transmission link between New York and California? Why?
8. Briefly describe the three modes of transmission.
9. What are the three categories of bandwidth? Which one would be used for video transmission?
10. Describe the differences between amplitude and frequency modulation.
11. What is multiplexing and what purpose does it serve?

References and Bibliography

Astrain, S. "INTELSAT and the Digital Communication Revolution." *Telecommunications Policy,* September 1983, 181–89.

Bell, Trudy E. "Long Distance Fiber-Optic Networks, Direct-Broadcast Satellites, and Low-Cost PBXs Bring Increased Communications Capacity to Customers." *IEEE Spectrum,* January 1984, 53–57.

Bucsko, Janis K. "PABX/LAN Integration." *Infosystems,* March 1984, 78–82.

Dahod, Ashraf M. "Local Network Standards: No Utopia." *Data Communications,* March 1983, 173–80.

Freeman, Roger L. *Telecommunication Transmission Handbook,* 2d ed. New York: John Wiley & Sons, 1981.

Friedman, Selma. "Fiber Optics in LANs (Local Area Networks)." *MIS Week,* Vol. 4, No. 28 (July 13, 1983), 1, 10, 14.

Holtzman, Henry. "Local Area Networks Form Information Web." *Modern Office Procedures,* June 1982, 66–80.

Hunter, John J. "The Anatomy and Application of the T1 Multiplexer." *Data Communications,* March 1984, 183–95.

Jones, J. Richard. "Emerging Trends in Local Area Networks." *Telecommunications,* December 1983, 54–60.

Maddox, Brenda. *Beyond Babel.* New York: Simon and Schuster, 1972.

Martin, James T. *Telecommunications and the Computer,* 2d ed. Englewood Cliffs, N.J.: Prentice-Hall, Inc., 1977.

Moncalvo, A., and R. Pietroiusti. "Transmission Systems Using Optical Fibres." *Telecommunication Journal,* Vol. 49, No. 11 (February 1982), 84–92.

Morant, Adrian J. "Fiber Optics—Ready or Not?" *Telephone Engineering and Management,* May 15, 1983, 70–72.

Netschert, Bruce C. "The Bypass Threat—And What to Do About It." *Telephony,* July 18, 1983, 112–16.

Rutkowski, A. M. "The Impact of New Technology on Satellite Radiocommunication." *Telecommunications,* February 1983, 46, 47, 62.

Sanders, R. W. "Microwave Keeps Up With the Times." *Telephone Engineer and Management,* July 1, 1983, 72–73.

Schawlow, Arthur. "Advances in Laser Technology Bringing Growing Array of Communications Uses." *Communications News,* November 1982, 42–43.

Technical Staff, Bell Telephone Laboratories, Incorporated. *Transmission Systems for Communications.* Murray Hill, N.J.: Bell Telephone Laboratories, Incorporated, 1982.

Wiley, Joe M. "Networking Horizons with Protocol Converters." *Data Communications,* July 1983, 127–33.

6 Data Processing and Communications

The first union of computing and communications occurred in 1940 when Dr. George Stibitz demonstrated the use of an electrically operated digital computer developed by Bell Laboratories. Stibitz, a mathematician, took a teletypewriter to a meeting of the American Mathematical Society at Dartmouth College in Hanover, New Hampshire. The teletypewriter in New Hampshire was connected to the calculator in New York via telephone lines. Stibitz then invited conferees to type in problems; the machine sent back the answers in minutes.

In its broadest sense, data communications describes any method of moving data from one location to another. Data can be communicated by physical transportation—messengers, trucks, or airplanes—or by electrical means—input and output devices, electrical transmission links, and associated switching equipment. In many cases, the best way to get information from one place to another is by mail; this method is usually the most economical. However, when rapid transmission of information is important, electrical methods are required.

One of the first business applications of data communications was the Sabre airline reservation system, developed jointly by IBM and American Airlines. After six years of development, the system began operation in 1962. The system used telephone lines to link terminals placed throughout the United States with a central computer located in Tarrytown, New York.

Business and industry are increasingly recognizing the advantages of data networks. Although only about one percent of the computers sold in 1965 were linked to a data communication system, virtually all computers sold or leased in the United States today have communications capabilities.

Early History and Evolution of Data Communications

The earliest form of data communications was the *teletypewriter (TTY)*, a printing telegraph instrument having a signal-actuated mechanism for automatically receiving printed messages. (*Teleprinter* describes a receive-only unit that has no keyboard.)

Teletypewriter systems, which transmit data at the rate of ten characters per second, have been in use for many years and are still useful for certain applications. However, the communication lines that carried the teletype signals were capable of transmitting information at much faster speeds than the teletype machines were capable of sending and receiving. This condition, coupled with the development of electronic computers with their high-speed processing capabilities, focused attention on the need for faster methods of data communications.

Data Transmission Research

In an effort to solve this problem, the Bell System investigated the possible use of the telephone-switched network for data transmission. Since the telephone-switched networks had been constructed for analog voice transmission and computers used digital transmission, extensive testing was required to determine the feasibility of using the telephone networks to transmit data.

The results of this research were published by the Bell System in 1960 in a report entitled *Capabilities of the Telephone Network for Data Transmission*. It showed that the telephone-switched network could be used satisfactorily for data transmission. This was a most important finding, for had the telephone network proved unsatisfactory, it would have been necessary to build a new national network to accommodate data communications. The time and expense involved in such a gigantic undertaking would have been quite a deterrent to the growth of data communications. Instead, this discovery began an era of rapid growth in data communications.

Industry Growth

A number of factors have influenced the rapid growth of data communications over the last two decades:

1. The trend of business organizations has been toward decentralization, mergers, and conglomerates. As companies grew larger and became more geographically dispersed, the problems of internal communications increased significantly.
2. High-speed communications equipment has been developed, and standard communication facilities and flexible common-carrier services have become available. As technological development

resulted in sophisticated new communications equipment, businesses discovered new applications for its use.

3. Costs of electronic communications equipment have lowered dramatically. Advances in chip technology and integrated circuitry have resulted in greatly improved microprocessor capabilities at reduced costs.
4. The advent of competition in the communications marketplace has resulted in a wider variety of service offerings and competitive pricing.

Definitions and Basic Concepts

Data communications evolved from the union of communications technology and computer technology. The integration of these two makes it possible to transmit data to computers from remote locations.

The integration has also brought together professionals from both disciplines, each of whom viewed data communications from the standpoint of his or her training and experience. These divergent viewpoints have resulted in inconsistencies in terminology; there are no standard definitions for many of the terms used in data communications. Accordingly, the authors' definitions presented here are intended to clarify the meanings of the terms used in this text. The definitions were formulated by checking a number of different references, including the Consultative Committee for International Telephony and Telegraphy, the American National Standards Institute, and the International Communications Association (ICA).

Data communications can be defined as the movement of encoded data from one point to another by means of electrical transmission systems, including radio and optics. In this sense, a data communication system must have two characteristics:

1. The data is translated into a special code for transmission.
2. The translated code is transmitted by electronic means.

Data communications may be distinguished from telegraphy chiefly by the fact that some form of processing is involved either prior to or after transmission.

The term *data* refers to any representation, such as numbers, letters, or facts, to which meaning can be ascribed. The statistician uses the term *raw data* to describe unprocessed data, thus distinguishing it from *information*, which connotes processed or meaningful data. In this sense, what is information to one person may not be information to another.

Although there is a subtle semantic difference between the terms *data* and *information*, an electrical transmission system cannot differentiate between them. Also, since the purpose of transmitting data is to supply it to the receiver to whom it will be meaningful, the terms *data* and *information* can be considered synonymous from a telecommunications standpoint. They will be used interchangeably in this text.

Data communications is an integral part of a data processing system and, like voice communications, is a subsystem of the field of telecommunications. In fact, data communications is sometimes defined as "the portion of a telecommunication system concerned with transferring data."

Data Communication Systems

In the data processing cycle, data must be collected and moved to the processing unit before it can be processed; and before processed data can be used, it must be delivered to the user.

In the early days of the computer, a user had to be physically present at the computer site to access computer power, and data had to be physically transported from one site to another. However, telecommunication channels have made it possible for data to be transmitted electronically, thus servicing users at geographically dispersed locations.

Data communication systems, or *networks*, are designed to transmit data from one location to another electronically. The objective of data transmission systems is to provide faster information flow by reducing the time spent in collecting and distributing data. Data communication networks facilitate more efficient use of central computers by providing *message switching* capabilities. *Message switching* is the routing of messages among three or more locations using either *circuit switching* or *store-and-forward* techniques. If a telecommunication line is available, message switching is accomplished by instantaneous circuit switching. If all telecommunication lines are in use, store-and-forward procedures are used; messages are accepted and stored in the computer memory until a telecommunication line becomes available, then forwarded to the next location.

Teleprocessing

Teleprocessing is a form of information handling wherein data processing equipment is used in combination with telecommunications facilities. It includes both the transferring of data from one location to another and the processing of data. Data communications is an integral part of teleprocessing. (The term *teleprocessing* was originally an IBM trademark; however, this is no longer the case.)

Data Processing Configurations

All computer systems contain hardware for data input, central processing, and data output. They all use stored programs to perform basic operations. Their main differences are in size, storage capacity, processing speed, and cost. *Microcomputers* are the smallest general-purpose computers; they are often a special-purpose or single-function computer on a single chip, but they may perform the same operations as a general-purpose computer. *Minicomputers* are a medium-sized class of computers. They are larger and more expensive than microcomputers but smaller and less expensive than mainframes. Large computers, known as *mainframe computers*, are the backbone of the computer industry. They are capable of processing large amounts of data with very fast processing speeds.

Data communications is combined with data processing into two types of systems:

1. centralized systems, wherein the processing of data is done on a single computer
2. distributed systems, wherein the processing of data is centralized on more than one computer

The earliest data processing configuration was *centralized processing*, wherein the processing for several divisions, functional units, or departments is centralized on a single computer. In this configuration, the input/output devices are located in the same area as the computer. The data input and output are in the form of physical media, such as punched cards and printed reports. Early computers were designed to handle one task at a time—an expensive mode of operation—and were centralized in a company because computers were very expensive and their operation required a special environment and operators with special technical skills. Thus, it was not economically justifiable to have computers in several locations within the organization.

The next development was the teleprocessing system, wherein the components of the computer system are geographically separated but joined into a system by telecommunication channels. Modern computer systems permit the attachment of a variety of peripheral devices, such as keyboard input devices, card readers, and line printers. Human operators type data into *terminals*, devices with typewriter-like keyboards that are placed in remote locations. These remote terminals are linked to the computer by telecommunication channels, usually leased telephone lines.

In teleprocessing systems, as in centralized processing, the processing of data is done on a single computer. However, terminals located outside the data processing center are connected to the computer via telecommunication lines that transmit data to the computer for processing. The principal advantage of teleprocessing is the time savings in data transmission. Electronic methods of data transmission are considerably faster than physical methods such as hand delivering punched cards or tapes. Also, when data is keyed directly into terminals, the preparation of source documents is eliminated.

The third and most advanced electronic data processing configuration, *distributed processing*, allows both computers and the data required for processing to be distributed throughout the organization. The data communication network is incorporated as an integral part of the system; however, the essential characteristic of distributed processing is that more than one computer processes the data. Distributed processing systems have the capability to process data at multiple points within a network. If the capability to process the information at one site does not exist, the data can be moved quickly to another site where it can be processed.

A distributed system may contain microcomputers, minicomputers, or even larger general-purpose computers, referred to as host computers, located in the department where processing is required. These machines perform local processing tasks and transfer data between computers and terminals rapidly and efficiently.

Distributed processing has several advantages:

1. By using smaller computers to perform some of the processing tasks, less complex (and less costly) equipment is required at the centrally located computer. This generally reduces costs for system design and programming.
2. Dividing a complex computer activity into simple components increases system reliability, because failure of any component will not render the entire system inoperative.
3. When processing is moved toward the user, each branch has greater control over its own data processing functions.
4. There is less traffic on communication lines, resulting in decreased communication costs.
5. Since data is entered into the system at a local site rather than a remote site, transactions may be processed almost as soon as they are entered, providing system users with more timely information.

Because of the many advantages of distributed processing, there is a growing trend toward decentralizing data processing activities.

Data Transmission Systems

Data transmission systems are designed to serve a variety of applications; thus, they differ in the way they function. The type of system used is determined by the information requirements of the user, including the following factors:

1. the quantity of data to be transmitted
2. whether immediate action is required
3. *response time* (the interval between data input and the system's response to the input)
4. *delivery time* (the time from the start of transmission at the transmitting terminal to the completion of reception at the receiving terminal, where data is flowing in only one direction)
5. the kinds of input data and the accumulation process employed
6. the geographical location to the users of the system

Offline/Online Systems

Data transmission systems can be classified into two broad categories: offline and online. Online systems may be further divided into interactive and noninteractive systems.

Offline means that the data is not transmitted directly to the computer but is stored on magnetic tape or cards for later processing. Offline systems are *noninteractive* (that is, no interaction takes place between the user at a terminal and the computer during the execution of a program), since the data storage location is not directly connected to the computer.

Most offline systems employ *batch-processing* techniques: the input records are collected in their original, physical form (time cards, invoices, report cards), accumulated over a period of time, and transcribed onto an input medium that can be read by the computer. The records are then transported to the computer room in groups, or batches, and read into the main computer storage. They are processed in batches, and the output is transmitted to a designated storage area, terminal, or printer as a batch.

In remote batch processing, called *remote job entry (RJE),* data is collected at remote locations and transmitted periodically via remote input/output terminals over telecommunication lines to a centralized computer system. The data can be transmitted offline to an auxiliary storage unit and held for input to the computer at a scheduled time. Once processed, the output is transmitted back to the user located at the same remote terminal.

Batch processing is particularly appropriate for accounting applications, such as payroll. In this application, the employee time cards are gathered periodically and serve as the source documents for the preparation of the input media. The input records are then entered into the computer system for processing. In processing, the current input records are matched with the employees' permanent records; processing produces updated employees' cumulative records, pay drafts, and other required records.

Offline batch processing is comparatively inexpensive. It is used when current, up-to-the-minute information is not required. The predictable nature of batch-processing requirements permits optimal scheduling of computer time; thus, it is efficient and economical for use in recurring, routine applications.

Offline transmission systems were common in the early days of data communications; however, the present trend is to online systems. *Online* means that devices or subsystems are directly connected to the computer, and data flows directly between the terminal and the computer.

The transmission of online communications may be either batched or *realtime*. In realtime computing, processing is performed as the operator keys in the data; thus, output is received quickly enough to affect decision making. The fast response time of online realtime systems (OLRT) is a definite advantage. However, these systems are expensive to implement and to operate; thus, some applications employ batch-processing techniques even though input devices are online.

In an online system for batch processing, data is accumulated into groups, or batches, at a remote terminal. The data is transcribed from source documents onto an input medium (paper tape, magnetic tape or disks, punched cards, or computer memory) and sent over telecommunication lines in batches to the computer system. There it is processed as it is received, or it may be merged with input from other sources before processing. As discussed previously, remote job entry can be used with offline transmission systems. However, its most frequent use is with online batch processing. In online RJE processing, both computer programs and input data can be entered at remote locations and transmitted online to the central computer for processing. The output is received at the RJE location.

Both batch processing and RJE are noninteractive; the computer merely receives a batch transmission and responds to it on a programmed basis. Each RJE station simulates a computer system in itself, even though one control computer does the processing for more than one user. The use of RJE procedures permits users to share a computer's time.

Online Realtime Transmission Systems

Realtime systems are online telecommunication systems that provide immediate, two-way communication between terminals and a computer, processing transactions as they occur. The term *realtime* generally suggests fast response. The transmission occurs sufficiently fast that it can be used as if it were instantaneous.

Realtime systems are *interactive* because they allow the user to communicate directly with the computer as the user wishes during processing. They are described as transaction-oriented, since data is sent in single transactions rather than in batches. The data is entered directly into the main computer for immediate processing by the computer program. Each transaction is processed individually and results returned immediately. Then another message is sent, and the process is repeated.

Online realtime (OLRT) transmission systems employ the most advanced technology and the most sophisticated programming techniques in data communications today. These systems are characterized by their fast response; when information is needed quickly and is useful only when it is obtained almost immediately, a realtime system is required. An important advantage of realtime systems is that applications are possible that depend on the fast response time these systems provide.

Some jobs are not suited to batch-processing techniques because the data would become outdated before it was processed. For example, when a charge account customer requests the current account balance shortly after making a payment, the response supplied by a batch-processing system would be inaccurate since it would not reflect the recent payment transaction. Similarly, reservation systems used by airlines, railroads, hotels, and motels require constant updating as events occur in order to be useful (Figures 6.1 and 6.2). Thus, the current status of any reservation system can only be obtained by using a realtime processing system.

Multiprogramming An essential part of OLRT systems is their multiprogramming and multiprocessing capabilities. *Multiprogramming* means that two or more programs can be run simultaneously by interleaving their operations.

A computer system has the capacity to hold more than one program in storage at one time, but the *central processing unit (CPU)* can only process one instruction at any one time. Processing functions are executed at high speeds, but input and output operations are relatively slow. Since the input and output operations cannot keep pace with the processing operation, the CPU will be left idle much of the time. To reduce the time that the CPU is idle, realtime multiprogramming systems overlap input and output with processing. Thus, although the actual processing is performed sequentially, the system appears to be processing two or more programs at a given time in the same CPU. For example, in an interactive, overlapped terminal system an operator keys in a record and the system pro-

Figure 6.1
Airline Reservation Center

(Courtesy of United Airlines)

cesses and prints it—all so rapidly that the functions appear to be performed at the same time.

With multiprogramming, two or more independent jobs may be processed concurrently. Instead of allowing each job to run the complete cycle before it begins the next job, the computer switches back and forth between them, thus using the CPU to its fullest. The concept of multiprogramming can be extended to users at distant locations by using data communication channels. Multiprogramming is especially useful here because a great deal of the time it takes to run a data communication program on a computer system is spent in input and output operations. The use of multiprogramming allows a computer system to execute other application programs in the same time frame. Figure 6.3 illustrates how multiprogramming uses overlapped and nonoverlapped processing.

Multiprocessing Another feature of OLRT systems is *multiprocessing*, wherein two or more CPUs are interconnected into a single system and one control program operates both processors. In addition to providing faster processing, multiprocessing systems can be set up to allow one processor to take over if another fails. However, the use of a second computer on a standby basis alone is not multiprocessing. Multiprocessing depends upon the computers being

Figure 6.2
Airline Reservation Center

(Courtesy of United Airlines)

Figure 6.3
Overlapped and Nonoverlapped Processing

Nonoverlapped Processing

Input	-1-			-2-			-3-		
Process		-1-			-2-			-3-	
Output			-1-			-2-			-3-

Overlapped Processing

Input	-1-	-2-	-3-	-4-	-5-	-6-	-7-	-8-	-9-
Process		-1-	-2-	-3-	-4-	-5-	-6-	-7-	-8-
Output			-1-	-2-	-3-	-4-	-5-	-6-	-7-
Time Intervals	1	2	3	4	5	6	7	8	9

interconnected and both systems being under the control of a common operating program.

Multiprocessing is one of today's most sophisticated data processing techniques. It is used when the computing power of a single CPU is not sufficient to process the jobs to be done within a given time. Multiprocessing systems are useful for applications in which time is a critical factor, such as in airline reservations, intensive-care patient monitoring, and space flights.

Large computers in a multiprocessing system can also support multiprogramming. Each processor can function independently as a single computer and process two or more jobs nearly concurrently. Multiprogramming and multiprocessing increase processing speed and power and the processing efficiency of the computer system. They are widely used in business today because their capabilities permit work to be done that could not be accomplished at all without them.

OLRT—An Application

The development and implementation of online realtime processing systems can make a substantial contribution toward improving services and reducing costs in many industries. (See Figure 6.4.) One such example is how the telephone company converted its directory-assistance (formerly called "information") service to a realtime processing system.

For many years, directory assistance relied upon printed records. These records required daily updating as phones were installed and removed. The updating process was performed by company personnel using manual procedures. When a customer called directory assistance to request a telephone number, the operator searched for the number in a printed record (telephone directory). The operator had to flip pages constantly—a cumbersome, repetitive activity. Although the process was boring, it required a high degree of alertness and mental concentration to ensure speed and accuracy in output.

Now, using modern search techniques on a realtime basis, operators can search the computer files for number records by keyboarding a few identifying characters. The computer selects a few potential listings from which the operator can make a decision and relay the information to the customer who is waiting on the line.

For example, to search the computer files for the name ROBERT J. SMITH, 24708 HARRISON, DETROIT, MICHIGAN, the operator would key in several letters of the subscriber's last name, such as SMI; tab over to another column and key in one or two digits of the house or apartment number, such as 24; tab over to still another column and type the first letter of the street name, H. This would produce all the listings for all the subscribers whose last name started with SMI, whose house or apartment number began with 24, and whose street name commenced with H—perhaps ten or twelve in all. The operator would then evaluate these listings in terms of the

Figure 6.4
Computer Applications Using OLRT Systems

Application	Activity
Airlines	Flight schedules Reservations/cancellations
Artificial Intelligence	Computer games Linguistics (language translation) Robotics Simulating human dialogue
Banking	Account transactions Account inquiry Electronic funds transfer (EFT)
Education	Computer-Assisted Instruction (CAI) Student records Student registration Testing-scoring/distribution statistics
Health	Patient records/case histories Patient-care monitoring Diagnostic procedures Hospital information systems Computer-Assisted Research (CAR) Computer-Assisted Pharmacy Systems (CAPS) Psychoanalysis/therapy
Insurance	Policyholder information/records
Investments	Market quotations/transactions Market research Financial analysis/planning
Legal	Drivers' licenses and records National Crime Information Center Attorney business records Scheduling attorney time/cases/courtroom Legal research
Manufacturing	Computer-Assisted Design (CAD) Computer-Assisted Manufacture (CAM)
Politics	Computer voting systems Election return reports Fund raising Legislative administration
Retail Distribution	Point-of-sale (POS) systems Credit card authorization Inventory control Order processing Account records Electronic checkout
Science	Scientific research Scientific problem solving

requested information and make a decision. In a rare instance the operator might have to key in additional identifying information or even ask the customer for further information.

The use of realtime processing in this type of application has had two very desirable side effects:

1. The average time of a directory-assistance call has been cut from 36 seconds to 25 seconds, a reduction of over 30 percent.
2. The job is now more interesting because the boredom of manual procedures has been eliminated.

Reducing the average time for directory-assistance calls from 36 seconds to 25 seconds may not seem like very much. However, since the time the operator spends in learning what number the customer wants and the time the operator spends in relaying the number to the customer cannot be reduced appreciably, the chief candidate for possible time reduction is the actual search time. (The next time you make a directory-assistance call, notice how quickly the operator severs the connection after your acknowledgment of the information.) Given the large volume of directory assistance calls the telephone companies handle daily, a reduction of over 30 percent represents a significant saving in operating costs.

Job satisfaction is somewhat more difficult to measure but, nevertheless, very real. Using manual procedures, an operator is required to remain mentally alert while performing repetitive tasks at high speed and under the stress of time. Replacing manual procedures with computerized procedures makes the job more interesting and less tiring. The intangible benefits to the workers can have a positive effect upon employee morale and productivity.

Combined Processing Systems

Many computer systems are capable of both OLRT and batch-processing operations. The same data base and the same online terminals used for realtime processing can be used to supply input data for batch processing. Many applications use both realtime processing and batch processing at various stages in their processing cycle. Realtime processing is used when fast, two-way interaction is required; when these conditions are not required, batch processing is satisfactory and results in cost savings.

In the previous example of telephone company directory assistance calls, OLRT processing was necessary to provide the information requested while the customer waited on the line. However, the same data base used to provide this information is also used for compiling and printing an annual telephone directory. In this case, the data base is read onto magnetic tape and processed using batch techniques. The tape is then sent to the printer who uses it to automatically print the telephone directory.

Many business applications that use OLRT systems also employ batch-processing techniques at some stage of their processing. The investment business provides another example of the use of combined systems. Brokerage firms use realtime systems for stock price quotations, research updates, trading transactions, and account information that must be up to the minute. Batch-processing techniques are used for daily summaries, closing prices, customer statements, and various financial reports that are less dependent on time.

Time Sharing

Time sharing allows several computer users to share the facility for different purposes on what appears to be a simultaneous basis. A time-sharing system is an online realtime system that uses terminals and communication networks and employs multiprogramming techniques. Multiprogramming allows electronic data processing systems to handle more than one program at any one time; similarly, time sharing allows the EDP system to serve more than one user at any one time.

A time-sharing system includes relatively slow-speed terminals that are online to high-speed, direct-access storage devices and a CPU. Because CPU execution time is so much faster than operator reaction time, the processing unit can be switched from one terminal to another and give almost immediate responses to each user. The users are seldom aware that the system is being shared. Instead, it is as if each user has sole access to the computer.

Time sharing assumes that no one user will have enough work to keep a computer busy, given its tremendous electronic speed. It reduces computer idle time, thereby making more efficient use of computer resources.

In a time-sharing system, the users share computer power and computer costs. Three costs are involved:

1. the cost of the terminal
2. charges for the use of communication lines
3. charges for the use of the computer itself

The terminals may be either purchased or leased. Charges for the use of the telephone lines to connect the terminal to the computer depend upon the distance from the terminal to the computer, the length of time the terminal is used, and the time of day the telephone lines are used. The charge for the computer facility is based on the actual time the computer is used, calculated in seconds, minutes, or hours.

The advantage of using a time-sharing system is that the user is charged for the telephone lines and the central processing facilities only for the actual time they are used. This permits users to access a computer system at a fraction of the cost of installing their own system and provides many users with computer services they could not otherwise afford.

There are two basic approaches to the use of time-sharing facilities. In the first approach, an organization purchases and operates its own time-sharing system for its own use. This is known as an *inhouse system*; it enables several departments within the organization to share one computer.

In the second approach, time-sharing services are rented from a commercial firm, usually known as a time-sharing *service bureau* or an *information utility*. These businesses rent computer time and realize their profits from high utilization of their processing equipment. A check of the telephone directory in major cities under the heading "Data Processing" reveals numerous listings in this highly competitive field. To obtain a marketing advantage, some service bureaus are now offering additional features, such as sophisticated programs and extensive data banks for customer use. In addition, some service bureaus will complete an entire job for their customers.

The two principal forms of time-sharing processing are conversational (interactive) and remote batch processing.

Conversational Time-Sharing Processing In the conversational mode, the user may retrieve programs previously written and stored in the computer, or the user may write a new program and enter it into the computer for processing. Almost any computer language can be used on a time-sharing network. One of the most popular time-sharing languages is BASIC (Beginners' All-purpose Instruction Code), a language created for use in time-sharing systems.

BASIC is usually implemented as an interactive language; that is, it permits interaction between the user and the computer while a program is being executed. Most versions of BASIC lead the user through the program step by step, providing instructions after each statement and pointing out any errors in the statement.

An important advantage of BASIC is that it is easy to learn. It was designed for the user with very little computer knowledge who wants to use the computer for solving problems. Because of its simplicity, BASIC has grown rapidly in popularity.

Remote Batch Time-Sharing Processing Remote batch timesharing is the shared use of a central computer by many users with data being processed by batch techniques. Remote batch time-sharing processing has the same features and limitations as any batch system: i.e., there may be a time lag of hours, days, or even weeks in obtaining responses.

The possible applications of conversational time sharing and remote batch time-sharing are identical to those being processed without the time-sharing feature.

System Justification

The costs of designing, implementing, and operating a realtime processing system are substantially greater than those for a batch system. In evaluating a realtime system, the additional costs must be weighed against the increased benefits it provides. Frequently, however, the benefits are intangible or hard to measure, such as improved customer service or more timely information.

A potential hazard in system planning is the tendency to overdesign. Faced with a wide variety of system features from which to choose, it is tempting to incorporate the "nice-to-have" features rather than concentrating on basic requirements.

Summary

A modern data communication system has two characteristics:

1. Data is translated into a special code for transmission.
2. The coded data is transmitted by electronic means.

Data transmission systems provide faster information flow by reducing the time spent in collecting and distributing data.

The phenomenal growth in data communications can be attributed in part to the fact that the telephone-switched network can be used for data transmission. The digital signals used in computers are converted to analog signals for transmission and reconverted to digital for computer processing.

Teleprocessing means data processing at a distance. It includes both the transferring of data from one location to another and the processing of data. In teleprocessing systems, the processing of data is done on a single computer; in distributed systems, data is processed on more than one computer. In distributed processing systems both computers and the data required for processing are distributed throughout the organization. Distributed systems have the capability to process data at multiple points within a network.

Data transmission may be classified into two broad categories: offline and online. Offline means that the data is not transmitted directly to the computer but is stored on magnetic tape, disks, or cards for later processing. Online devices are connected directly to the computer, and the data is sent directly between the terminal and the computer. Realtime systems are online systems that provide immediate, two-way communication between terminal and computer and process transactions as they occur. Online realtime systems make possible many processing applications that could not be performed without the fast response these systems provide. For example, reservation systems used by airlines, hotels, and motels require constant updating as events occur.

An essential part of OLRT systems is their multiprogramming and multiprocessing capabilities. Multiprogramming means that two or more programs are executed simultaneously by interleaving their operations. This technique permits two or more independent jobs to be processed concurrently. In multiprocessing, two or more CPUs are interconnected into a single system, and one control program operates both processors. It is used when the computing power of a single CPU is not sufficient to process the jobs to be done within a given time. Large computers in a multiprocessing system can also support multiprogramming. Each processor can function independently as a single computer and process two or more jobs concurrently. Multiprocessing is one of the most sophisticated data processing techniques available today.

Review Questions

1. What are the two identifying characteristics of data communication systems?
2. What is the objective of data communication systems?
3. Distinguish between data communications and teleprocessing.
4. Distinguish between offline and online operations.
5. What is the principal advantage of online realtime processing? What is the principal deterrent to its use?
6. When is it appropriate to use offline batch processing? What is its principal advantage?
7. Distinguish between multiprogramming and multiprocessing.
8. Describe the concept of time sharing. When is its use appropriate?

References and Bibliography

Alexander, A. A., R. M. Gryb, and D. W. Nast. "Capabilities of the Telephone Network for Data Transmission." *Bell System Technical Journal,* 39, May 1960, 431–76.

Breslin, Judson, and C. Bradley Tashenberg. *Distributed Processing Systems.* New York: AMACOM, A Division of American Management Association, 1978.

Cypser, B. J. *Communications Architecture for Distributed Systems.* Reading, Mass.: Addison-Wesley Publishing Company, 1978.

Data Communications. *Executive Guide to Data Communications,* 5th Volume. New York: McGraw-Hill Publications Company, no date.

Davenport, William P. *Modern Data Communications.* Rochelle Park, N.J.: Hayden Book Company, Inc., 1971.

Dix, John. "AT&T Takes on the Computer Market." *Computerworld,* April 2, 1984, 1, 8.

FitzGerald, Jerry, and Tom S. Eason. *Fundamentals of Data Communications.* New York: John Wiley & Sons, Inc., 1978.

Gentle, Edgar C., Jr., ed. *Data Communications in Business.* New York: American Telephone & Telegraph Company, 1965.

Hindin, Harvey J. "What American Bell Offers." *Electronics,* June 30, 1982, 68–69.

Kroenke, David M. *Business Computer Systems,* 2d ed. Santa Cruz, Calif.: Mitchell Publishing, Inc., 1984.

LaBlanc, Robert E. with Richard M. Wolf and Elizabeth A. LeBlanc. "Communications + Data = Compunications." *Telephone Engineer and Management,* May 1, 1983, 67–71.

Mandell, Steven L. *Computers and Data Processing,* 2d ed. St. Paul, Minn.: West Publishing Company, 1982.

Martin, James T. *Telecommunications and the Computer,* 2d ed. Englewood Cliffs, N.J.: Prentice-Hall, Inc., 1977.

Pitt, Daniel A. "Interaction Between Voice and Data Elements in a Local Area Network." *Telephony,* March 7, 1983, 40–42.

Popkin, Gary S., and Arthur H. Pike. *Introduction to Data Processing,* 2d ed. Boston: Houghton Mifflin Company, 1981.

Sanders, Donald H. *Computers Today.* New York: McGraw-Hill Book Company, 1983.

Sardinas, Joseph L., Jr. *Computing Today.* Englewood Cliffs, N.J.: Prentice-Hall, Inc., 1981.

Shelly, Gary B., and Thomas J. Cashman. *Introduction to Computers and Data Processing.* Fullerton, Calif.: Anaheim Publishing Company, 1980.

Sherman, Kenneth. *Data Communications.* Reston, Va.: Reston Publishing Company, Inc., 1981.

Techo, Robert. *Data Communications.* New York: Plenum Press, 1980.

Data Communication Systems

7

The increasing use of computers in business operations has created the need for rapid movement of data from one location to another. New computer hardware using the latest technological developments requires a continuing evaluation of information movement procedures in order to maximize the operational efficiency of the computer system. The manager has two options in moving data: physical transportation of the documents or electrical transmission of the information. If there is no immediate need for the data, physical transportation could be the best alternative. If speed is a critical factor, electronic data transmission is required. The data transmission system should provide the same degree of speed and efficiency as the computer system to realize the full system benefits. To use an analogy: A chain is no stronger than its weakest link. The manager is responsible for providing a data communication process that maximizes the information-processing system and is not the weak link.

Data communication systems are networks of components and devices organized to transmit data from one location to another—usually from one computer or computer terminal to another. The data is transmitted in coded form over electrical transmission facilities.

This chapter outlines the basic hardware, controls, and procedures of data communications and describes how they function together as an integrated data communication system.

Components of a Data Communication System

The three essential components common to all data communication systems are the *source*, the originator of the information; the *medium*, the path through which information flows; and the *sink*, the receiver of the information.

Figure 7.1
A Simple Data Communication System

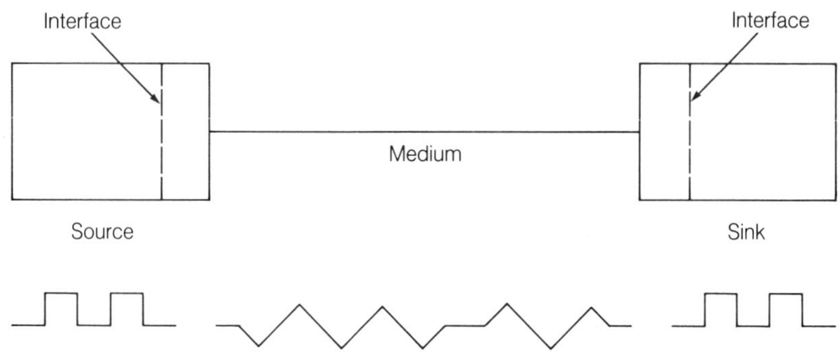

The source in a data communication system is *remote terminals,* usually devices with typewriter-like keyboards used for entering data. Remote terminals might be in a different department, a different building, or even a different geographic location from the mainframe or CPU. The medium in a data communication system is the communication link, the facility that links remote terminals to the CPU. The medium could be wire, radio, coaxial cable, microwave, satellite, or light beams. Leased or public telephone lines are the most frequently used communication medium. The sink in a data communication system is the computer system that processes the data received.

To these basic components we could add the *modem* that transforms input into transmittable signals. Two modems are required—one to convert data from the source into a form that can be transmitted over telephone lines, and the other to reverse the process at the terminating end. In two-way communication the source and the sink could be constantly changing roles; that is, the terminal might alternate as both a source and a sink.

In a simple data communication system, an example of which is shown in Figure 7.1, the source might be a terminal keyboard; the medium, a telephone line; and the sink, a computer. If the system operated in a query-response mode, the terminal and computer would change roles as the transaction progressed. Thus, the terminal, which was originally the source, would become the sink; and the computer, which was orginally the sink, would become the source.

Terminal Equipment

Originally, the word *terminal* meant the point at which data could enter and leave the communication network. In practice, however, it has taken on an additional meaning. The word *terminal* has become synonymous with *terminal equipment* and refers to any device capable of either input or output to the communication channel.

Terminals provide interfaces with computer systems so that people can insert or extract data; terminals also provide a convenient way for people to exchange data directly. Terminals receive input data in coded form and convert it into electrical signal pulses for transmission to the computer. Similarly, terminals at the receiving end transform the electrical impulses into characters that can be read by humans. Thus, terminals function as translation devices for communication codes. Data can be entered into terminals either by human operators or by machines that collect data automatically from recording instruments.

Prior to 1968 the major terminal device was the teletypewriter, and the principal input medium was punched paper tape. Each keystroke on a transmitting teletypewriter (source) produces a sequence of electrical pulses determined by the coding representation for the keyed character. The electrical signals are sent over the communication channel (medium) to a receiving teletypewriter (sink), where they are reconverted to their original form.

The Carterfone decision lifted the ban on the attachment of customer-provided terminal equipment to the telecommunication network. This decision, along with phenomenal advancement in the computer industry, promoted competition among terminal suppliers, and the terminal industry grew rapidly. Today, there are many different kinds of data communication terminals offered by many manufacturers. Thus, system designers can be very selective in their choice of equipment. However, the wide variety of terminals available can cause confusion and make selection difficult. An understanding of terminal characteristics and capabilities will be helpful in the selection process.

There are undoubtedly many ways to classify terminals. These might include such factors as operating speed, type of transmission (batch or online), memory capability, mode of operation (HDX, FDX), transmitting capability (send only, receive only, send/receive), method of input (keyboard, card reader, tape reader), output form (hard copy, soft copy), error control, type of applications, and, of course, price.

This chapter will discuss the following five broad categories of terminals:

- ☐ typewriter or keyboard terminals
- ☐ video terminals
- ☐ transaction terminals
- ☐ intelligent terminals
- ☐ specialized terminals

Figure 7.2
A Typewriter Keyboard Terminal

(Courtesy of United Airlines)

Typewriter or Keyboard Terminal This is the most widely used type of terminal. It resembles a typewriter and usually has a standard alphanumeric keyboard with special function keys to provide transmission control capability, such as the "line feed" or "bell signal." Typewriter terminals (Figure 7.2) print one character at a time and produce hard-copy printouts. Keyboard terminals are used on low-speed public or private telephone lines. They operate in two directions: sending and receiving. In the sending mode they are controlled by an operator; in the receiving mode they are controlled by either the central processing unit of a computer or by another operator at a distant machine. Typewriter terminals are usually unbuffered; that is, they have no storage capacity. Each character is printed on paper at both the sending and receiving ends as soon as it is keyed.

There are two types of keyboard terminals: those that are online continuously and those that are accessed on a dial-up basis. Both types of terminals may be *polled* (called up in sequence) by a central control unit to request the terminal to transmit a message.

Video Terminals Video terminals consist of a keyboard and a cathode ray tube (CRT), a visual display device resembling a television screen. The keyboard is the input medium; the operator can enter both the data and the control commands that direct the operation of the computer. The CRT provides *soft copy,* a visual display with no permanent record. Display terminals may also be obtained with a printer attached, thus enabling them to print hard copy.

Most video display terminals utilize a standard typewriter keyboard, plus control and special function keys, such as "insert," "delete," and "repeat." As the operator types in a character, it appears on the screen. These terminals have a *cursor* (from the Latin *cursus,* meaning place), which is usually a blinking symbol that indicates current location on the CRT screen. The operator can move the cursor horizontally and vertically to any desired position. An important advantage of CRT display terminals is their capability to edit text. Errors detected on the screen can be corrected, usually by backspacing and striking over the error. Entire words, lines, or paragraphs can be deleted or repositioned through the use of special function keys. Changes in input copy are possible because some amount of the copy is held in a *buffer,* or temporary, memory until the user presses a special function key to transfer it to the main memory. Video display terminals with alphanumeric keyboards have become well known through their use in word processing machines.

A special type of video display terminal is the graphics terminal that can display not only letters of the alphabet and numbers but also graphic images (charts, maps, and drawings). These terminals use matrix technology, in which many closely spaced dots are connected to draw lines and to plot data graphically. Graphic display terminals can accept input from a keyboard, an input tablet, and/or a light pen—an electronic drawing instrument equipped with a photoelectric cell at its end that allows the user to "draw" designs directly on the display screen. Some graphic display terminals are capable of displaying charts and drawings in different colors on the screen; the displayed material can also be reproduced as hard copy with a printer or plotter.

Scientists and engineers have long used graphics to represent information for review and analysis. Computer technology has extended the use of graphics to many other professions and industries as well. Computer graphics are currently being used in manufacturing to design automobiles, household appliances, and other products; in the health profession to assist in diagnostic procedures; in the transportation industry to design routes and schedules; and in the banking and investment industries to summarize and analyze financial data. They are also used to present business operating data that will help managers identity trends and relationships and make informed decisions. The adage "A picture is worth a thousand words" could be rephrased to "A graph or visual summary of data is worth many pages of computer printout."

Video displays are high-speed devices, since data output is not slowed down by being typed on paper. These terminals are

Figure 7.3
A Video Display Terminal

(Courtesy of United Airlines)

especially useful when output from a distant location is needed quickly. They are used by airline reservation systems (Figure 7.3) to determine flight space availability, by hotels and motels to determine room availability, by brokerage firms to transmit stock market quotations, and by insurance companies to access and update policyholder records.

Transaction Terminals These terminals are designed for use in a particular industry application such as banking, retail point-of-sale, or supermarket checkout.

In banking, transaction terminals are used to update both customers' passbooks and the bank's account records. Terminals are also used online for off-hours banking and for processing customer inquiries.

Retail point-of-sale terminals are used to record the details of the sale in machine readable form. Their functions generally include verifying credit, printing sales slips, maintaining a local record of transactions, and updating inventory control records. All of these functions, except credit verification, can be handled by offline terminals,

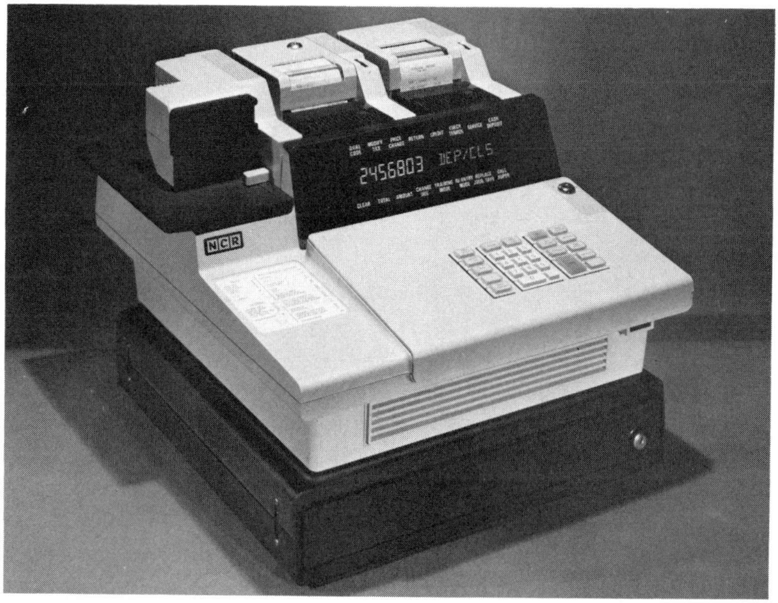

(Courtesy of NCR Corporation)

Figure 7.4
A Transaction Terminal

generally by cash register-type machines equipped with special keys to capture transaction data on paper or magnetic tape. Credit verification requires online access to storage files that may be built into the transaction terminal or may be in a central computer. See Figures 7.4 and 7.5.

Transaction terminals used in supermarket checkout lines have the capability to scan or read bar codes printed on the items being sold. See Figure 7.6. As the products go past the checkout point, the codes are read by a recording device or light pen that simultaneously prepares a cash register tape for the customer, records the sale, and updates the store's inventory. Transaction terminals are easy to operate and may be used by persons with little technical knowledge. They have become an integral part of business operations because they help to increase productivity and control costs. See Figure 7.7.

Intelligent Terminals Early types of terminals merely served as data input and output devices; they performed no processing, editing, or buffering. Such terminals are still used in certain applications, such as simple transaction recording. They are sometimes referred to as "dumb" terminals because of their limited capabilities. A teletypewriter is an example of a "dumb" terminal.

With the development of microprocessor technology, it became possible to incorporate some processing capability into peripheral devices, which greatly enhanced their usefulness. Computer

Figure 7.5
A Transaction Terminal

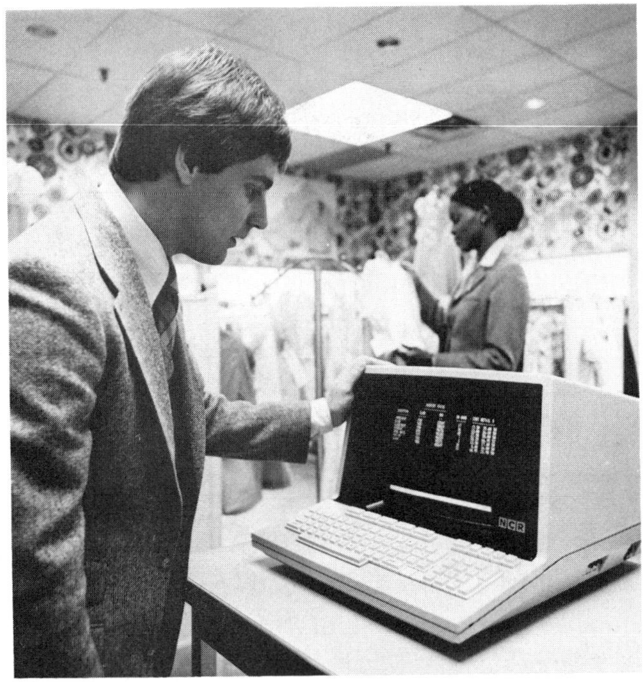

(Courtesy of NCR Corporation)

terminals equipped with a microprocessor are known as "intelligent" or "smart" terminals. They vary in degrees of enhancement. As additional intelligence is incorporated into these machines, more processing can take place at the terminal, relieving the burden on the mainframe computer. The more sophisticated terminals have become small computers in themselves, and they are frequently able to operate independently from the host computer.

Specialized Terminals Two of the newer types of terminals are audio response units and pushbutton telephones. Audio response terminals are unique in that their output or response is verbal rather than printed or visual. The input device may be either a keyboard or a telephone. In audio response units, the computer has a built-in synthesizer, enabling the computer to assemble prerecorded sounds into meaningful words. Unlike some mechanical voices that have low-fidelity, robot-like characteristics, the voice quality of audio response units is equivalent to that of the human voice. This synthesized response should not be confused with the response from an answering machine that transmits a recorded announcement.

The response from an audio terminal is designed for a specific type of message service. Messages are pieced together from sound frag-

(Courtesy of NCR Corporation)

Figure 7.6
NCR Supermarket Scanning System

(Courtesy of NCR Corporation)

Figure 7.7
NCR Lodging System

ments to produce a reply to a particular inquiry, such as a telephone number request from an information bureau. In this process an information operator finds the requested number and points to the appropriate number displayed on the CRT screen with a light wand. The operator is then released from the call, and the audio response unit synthesizes the message (in this case, the requested telephone number) and transmits it to the customer waiting on the line. The combination of human operators and audio response units saves human effort, thereby improving productivity.

Rotary-dial telephones are not generally used for data transmission since the rotary dial cannot be used as an input terminal. The dial is used to establish a call using dial pulses that are digital in form. However, once the connection has become established, the dial pulses are ineffective for signaling.

The pushbutton telephone has been used for some time in voice communications, but its use as an input terminal for data transmission is relatively new. Its widespread availability makes it particularly useful. A pushbutton telephone has a keyboard to replace the rotary dial; the keyboard is an integral part of the instrument. Pressing a key on the telephone set transmits a distinctive signal representing a number that the computer uses for the processing operation.

Many banking institutions offer a service that permits their customers to conduct certain banking transactions from their home or office using pushbutton telephones. For example, where this service is available, a customer can transfer funds from a savings account to a checking account and vice versa or pay bills to a selected list of merchants and utilities by keying a sequence of numbers representing a code into the telephone set. An interesting feature of these transactions is that while the customer communicates with the computer by pressing the appropriate pushbuttons on the telephone, the computer communicates with the customer by using an audio response unit.

Other Input/Output Devices There are many types of input and output devices. Figure 7.8 summarizes the principal ones. Frequently, telecommunication systems use a combination of these devices to perform a specific function.

Figure 7.9 summarizes the important characteristics of the five categories of data communication terminals.

Terminal Selection

The application dictates the type of terminal device required. Because of the wide variety of terminals available, the systems designer should understand the capabilities of the various types of terminals to match the capabilities with the applications.

Figure 7.8
Input/Output Devices

Document-Input Devices	Human-Input Devices	Input/Output Devices
Paper tape reader	Typewriter keyboard	Typewriter
Magnetic tape reader	Matrix keyboard	Printer
Punched card reader	Special keyboard	CRT screen
Magnetic card reader	Pushbutton telephone	Facsimile machine
Optical character reader (OCR)	Teleprinter	Magnetic tape or disk
Magnetic ink character reader (MICR)	Light pen	Passbook printer
Mark sense reader	Light wand	Transaction printer
Microfilm	Stylus	Audio response unit
Facsimile machine	Voice instruction	Light display
Tape cassette		Microfilm/fiche
Magnetic disk		Videotex
		Word processor
		Telephone (computerized voice)

Figure 7.9
Categories of Data Communication Terminals

Summary of the important characteristics of the five categories of data communication terminals.

- **Typewriter or keyboard terminals**
  - Hard copy output
  - Used on low-speed public or private lines
  - Both send and receive
  - Online or dial-up
  - Usually unbuffered
- **Video Display Terminals**
  - Feature a cathode ray tube (CRT)
  - Have editing capability
  - Buffer memory
  - Graphics display capability
  - Used on public or private lines
  - High-speed operation
- **Transaction terminals**
  - Used in business transactions
  - Designed for a specific application
  - Easy to operate
  - Usually buffered
  - Used on private lines
- **Intelligent terminals**
  - Feature processing capability
  - High-speed operation
  - Buffered
  - Used on public or private lines
  - Capable of operating independently of host CPU
- **Specialized terminals**
  - Audio response units—synthesized verbal output
  - Pushbutton telephones—pushbuttons provide input to computer

Some of the questions that might be considered in the selection process include:

1. Will the equipment perform the required function or functions satisfactorily?
2. Is the terminal response time consistent with efficient operation?
3. Is the capability for error correction built in, or will excessive operator attention be required?
4. Are the technical specifications (such as transmission code, operational mode, synchronization techniques) compatible with other units in the system?
5. Could the operation under consideration best be performed at the centralized processor or at a terminal?
6. What degree of intelligence is required? Is processing capability required? Is programmability required?
7. Can the terminal be upgraded to meet future needs?
8. Will the advantages of improved operations justify the cost of the equipment?

Modems

A basic problem of data communications has been sending digital signals over transmission facilities designed originally for analog voice transmission. When the communications line carries analog signals, an interface device is required. This function is performed by a *modem*, a combination of modulator and demodulator.

The modem and the communications line can be connected directly (hard wired) or indirectly (acoustic or inductive coupling). Acoustically coupled modems are portable; they can be used with any available telephone. These modems can transmit and receive when the telephone handset is coupled, or cradled, into the proper position in the modem. With acoustic coupling, the DC signals are converted to audible sounds, which are then picked up by the transmitter in an ordinary telephone handset (Figure 7.10). The audible signal is converted to electrical signals and transmitted over the telephone network. The process is reversed at the receiving end. Because they involve an extra conversion step (digital to audible to electrical) that can introduce noise and distortions, acoustic couplers generally are not as reliable as direct electrically connected modems. A direct connection to the communications line is preferable.

Modems come in a variety of shapes, sizes, and forms. They are often classified by speed; however, there is no consensus on the range of speeds in the various groups. Low-speed modems are sometimes defined as those that operate up to 1200 or 1800 bps (bits per second), medium-speed modems up to 4800 bps, and high-speed modems up to 9600 bps.

Figure 7.10
Components of a Data Communication System

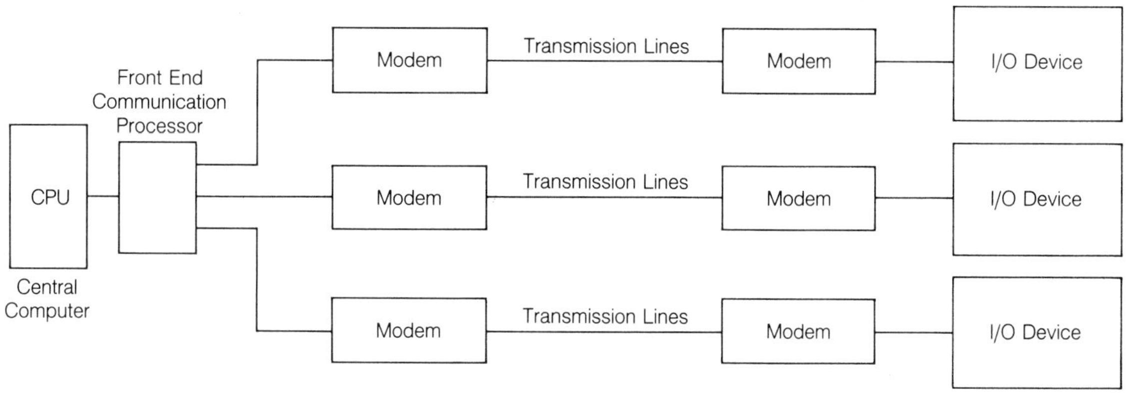

The previous discussion of modems concerned data communication over analog facilities. However, not all transmission facilities are analog. Many of the newer facilities are digital because of their ability to provide good, economical voice communications, as well as their ability to support data communications in digital form. Of course, where the transmission facility carries digital signals, no modem is required. AT&T's Digital Data System (DDS) uses digital transmission facilities, interconnected to form a synchronous network for data communications. The service became available in 1974 and is expected to grow to include over 100 metropolitan areas.

Front-End Processors

As terminals are added to a data processing system, the number and complexity of the operations necessary to handle them grow, and the demands on the CPU increase tremendously. To relieve some of these demands and to control the flow of data and ensure compatibility within the system, a *communications control unit (CCU)* or *front-end processor* is employed. This processor "front ends" a mainframe central processor by functioning as an auxiliary computer system that performs network control operations. This releases the central computer system to do data processing. Most front-end processors are minicomputers. The functional components of the front-end processor may be freestanding pieces of equipment or combinations of components integrated in one or more equipment units. Some of the functions performed by the front-end processors include:

1. Line access: connects communication lines to the main computer.
2. Line protocol: monitors line control procedures.
3. Code translation: translates the internal code of computer systems into communication codes.

4. Synchronization: ensures that the incoming signals are compatible with the requirements of the computer.
5. Polling: polls terminals to inquire whether they are ready to receive a message or whether they have a message to send.
6. Error control: checks accuracy of data received using parity techniques.
7. Determines path routing: chooses an alternate path to avoid heavy traffic or excessive error rate.
8. Flow control: controls the flow of the signal from the processing unit to its destination.

Central Processing Unit

The sink, the receiving unit in a data communication system, is generally a computer whose principal component is a central processing unit. The CPU is the heart of the computer system; it controls the input, data transfer, calculations, logic, and output operations of the system.

Transmission Links

The components of a data-communication system are connected by transmission links. These links could be wire, radio, coaxial cable, microwave, satellite, or light beams. Transmission links are discussed in more detail in Chapter 5.

Development of Data Codes

Data is transmitted over telecommunication facilities and entered into computers in a code based on binary digits.

The Language of Data

The binary numbering system is the language of data. The word *binary* (from the Latin *bini*, meaning two-by-two) indicates that two digits make up the system. A binary digit can be represented mathematically by a + or a − or a 1 or a 0. The + represents the flow of electricity; the − represents the absence of the flow of electricity.

In data communications, the smallest element of information is the *bit*, derived from the contraction of binary digit. Thus, each of the binary digits, 0 or 1, is referred to as a bit. In the same way that a bit represents a unit of information, a baud represents a unit of signaling speed. Each is characterized by the presence or absence of a pulse of electricity in a transmission channel. In a transmission

Figure 7.11
A Five-Channel Code

A	B	C	D	E	F	G	H	I	J	K	L
1	1	0	1	1	1	0	0	0	1	1	0
1	0	1	0	0	0	1	0	0	1	1	1
0	0	1	0	0	1	0	1	1	0	1	0
0	1	1	1	0	1	1	0	0	1	1	0
0	1	0	0	0	0	1	1	1	0	0	1

Magnetic Tape

system using binary representation, the term *baud* is synonymous with bits per second. However, in transmission systems using other than binary representation, baud and bits per second are not synonymous.

Some transmission systems combine groups of bits into *dibits*, a transmission concept wherein two bits are transmitted at a time, thereby doubling the transmission speed. Groups of bits may also be combined into other configurations, thereby departing from the binary concept. In these configurations, the terms baud and bits per second are not synonymous.

Basic Concepts of Codes

The American National Standards Institute (ANSI) defines a *code* as "any system of communication in which arbitrary groups of symbols represent units of plain text of varying length." Similarly, a *code set* is "the complete set of representations defined by a code," and *coding* is "the process of converting information into a form suitable for communications." For example, the representation of the characer *r* by a group of bits, such as 01010, is an example of coding.

The most frequently used medium for encoded data is magnetic tape. The presence of one type of magnetic mark represents the binary 1; the presence of another type of magnetic mark represents the binary 0. Magnetic marks are combined in various ways to construct codes.

Figure 7.11 illustrates a 5-channel code recorded on magnetic tape. The channels run the length of the tape, and each coded character occupies a column across the width of the tape.

To expand the number of characters that can be represented, 6-, 7-, and 8-channel codes have been developed. These tapes follow the same general pattern as the 5-channel tapes, except that a parity bit is often added in the last channel for error detection.

Figure 7.12
A Simple Telegraph

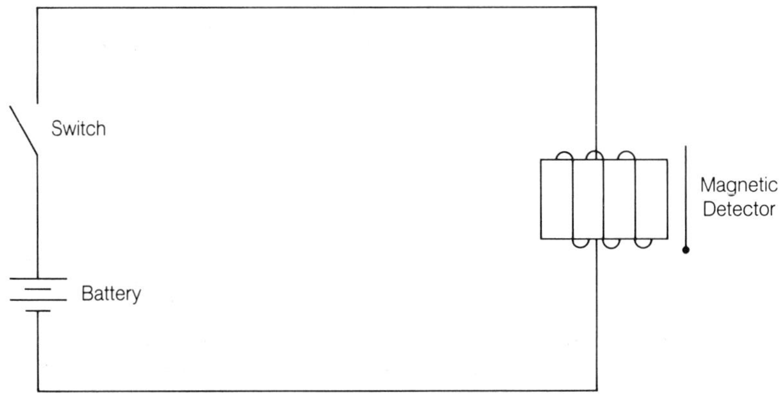

Voice Transmission of Data

Voice transmission over telecommunication facilities takes place in a conversational mode similar to face-to-face conversation. The words spoken into a telephone travel over an electronic connection to another telephone where they are reproduced in the same form and with the same characteristics as when they entered the communication system.

Data can be transmitted by voice over a telephone, but the speed and accuracy of the transmission is not satisfactory for most purposes. For example, sales representatives could verbally give hourly reports of sales volume to a central location from all sales locations. However, this method would be a very slow process, In addition, the collected data would be unreliable unless painstaking verification procedures were used. A more efficient method of transmitting data is to encode the message into a form that can be sent rapidly and accurately over telecommunication lines. To enable the data message to be understood, both the sending and receiving persons or machines must have the same understanding of coding details.

Origin of Codes

The development of codes used in data communications began with the invention of the telegraph by Samuel F. B. Morse in 1844. The clicks of the telegraph were unintelligible until a code was developed. The telegraph consisted of sending and receiving stations connected by an electrical circuit. The sending station contained a source of power and a switch that could be opened and closed to send pulses of electricity timed as dots and dashes (Figure 7.12). The receiving station contained a sensing unit that could detect the opening and closing of the switch. An electromagnet functioned as the sensing unit and detected the presence of a dot or a dash as transmitted by the sender.

The telegraph operator could send short or longer bursts of electric current by pressing down the switch for short or longer periods of time. A short burst of current represented a dot, and the longer burst (equal to three dots run together with no space between) represented a dash. Thus, the receiver could interpret each signal in sequential manner.

Data Transmission Controls

Data communication systems generally incorporate controls to:

1. specify the rules to be followed in transmitting data
2. detect errors in data transmission
3. ensure system compatibility

Protocols

As the options and intelligence available in data communication terminals increased significantly, technology mandated new rules and procedures for efficient operation of the system. These communication controls are known as *protocols;* they are formulated by the equipment suppliers. They may govern communication lines, types of service, modes of operation, circuit compatibility, or total networks. This is analogous to the use of traffic rules (highway ordinances) to control the efficient flow of traffic over city streets and highway networks.

Early protocols were referred to as *handshaking*. However, current usage of this term generally refers to the fact that a connection has been established and that the communication line is ready for the message. In present-day usage, protocol includes both handshaking and *line discipline,* a term that denotes the sequence of operations involving the actual transmitting and receiving of data. Many vendors use line discipline synonymously with protocol. Figure 7.13 illustrates line-polling protocol. Line polling is similar to a two-way conversation between a computer and a terminal wherein each confirms to the other the status of a message.

Error Detection

Morse code is used primarily for the transmission of information from one person to another. If the receiving operator fails to understand a signal or if a sending error has been made, it is easily

Figure 7.13
Line-Polling Protocol

Step	Computer		Terminals
1	Term #1, Any Msg?	--------------------	
2		--------------------	Nothing now
3	Term # 2, Any Msg?	--------------------	
4		--------------------	Yes
		--------------------	Here is Msg
		--------------------	Here is Parity Check
5	Parity is OK	--------------------	
6		--------------------	Here is More Msg
		--------------------	Here is Parity Check
7	Parity is OK	--------------------	
8		--------------------	End of Msg
9	Term #3, Any Msg?	--------------------	
10		--------------------	Nothing Now

discerned by either operator. It can be corrected by a retransmission of the questionable part. However, when data is sent between mechanical devices, there is a need for a systematic way to detect whether the message is valid or if something has gone wrong.

A basic problem in using voice facilities for data transmission is the presence of noise and distortion. The resulting errors in transmission require some error-control mechanism.

There are a number of methods of detecting errors. However, the most commonly used methods add redundance to the message to detect when a character is in error. A classical method of error detection is *parity checking,* which involves the use of a single bit, known as a *parity bit,* for the detection of errors. A description of parity checking will be helpful in understanding how error detection can be built into a coding structure.

Parity describes a condition wherein the total number of 1 bits in each character is always even or always odd, depending upon the parity system being used. When a parity system is being used, the transmitting equipment automatically adds one noninformation-carrying bit, called a parity bit or *check bit,* to the characters being transmitted (Figure 7.14). This enables the computer to run its own check on every character it processes.

There are two levels of parity error detection: character checking and block checking. The simplest and least expensive method of error checking is character checking odd or even parity. The parity bit that is added will be either a 0 bit or a 1 bit, whichever is required

Figure 7.14
Odd and Even Parity

Letter M	1	0	1	1	0	0	1	0	Even Parity Code
Letter J	0	1	0	1	0	0	1	0	Odd Parity Code

The eighth bit is the parity code.

Figure 7.15
Vertical and Longitudinal Redundancy Checking

Vertical Redundancy Checking

	Character 1	2	3	4	5
1	0	1	0	0	1
2	1	0	0	0	0
3	0	0	1	1	0
4	0	1	1	1	1
5	0	0	0	0	1
6	0	0	0	0	0
7	1	1	1	1	1
Odd Parity	1	0	0	0	1

Longitudinal Redundancy Checking

	Character 1	2	3	4	5	Block Parity
1	0	1	0	0	1	1
2	1	0	0	0	0	0
3	0	0	1	1	0	1
4	0	1	1	1	1	1
5	0	0	0	0	1	0
6	0	0	0	0	0	1
7	1	1	1	1	1	0
Odd Parity	1	0	0	0	1	1

to make the total number of 1 bits even or odd, depending upon the parity system being used. In a system using even parity, the total number of 1 bits, including the parity bit, should be even; in a system using odd parity, the total number of 1 bits, including the parity bit, should be odd. In a 7-bit code with even parity, the parity bit would be the eighth bit. If the character is represented by an even number of 1's, the parity bit added will be 0. If the character is represented by an odd number of 1's, the parity bit added will be 1. Checking consists of determining whether the data received conforms to the parity system being used. Character parity checking cannot detect errors involving the loss or addition of 2 bits in a character or compensating errors. For these types of errors, a longitudinal system known as *block character checking* or *longitudinal redundancy checking* (Figure 7.15) is employed.

Using longitudinal redundancy checking an additional 7 bits are added at the end of a specific block or message. Parity bits are added to all of the 1 bits, all of the 2 bits, all of the 3 bits, and so on until all of the bits have a bit at the longitudinal end, providing either even or odd parity at the end of the message. This checking significantly increases the probability of error detection.

Figure 7.16
Asynchronous and Synchronous Transmissions

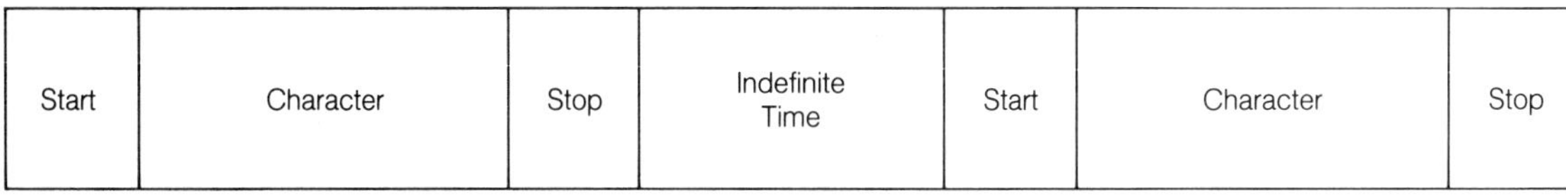

Asynchronous Transmission Each Character is an Independent Entity

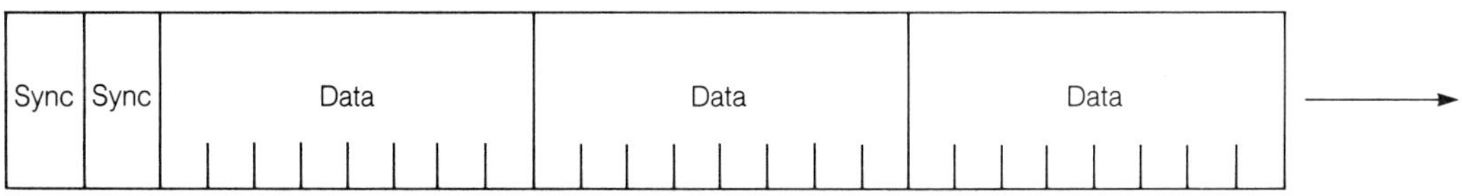

Synchronous Transmission Whole Blocks of Data are Transmitted After Synchronization

Synchronization

Data signals, or bits, are transmitted in a specific code and transmission sequence. It is important that the coding and sequencing be sent properly so that devices that receive the data signals will be able to interpret them as intended. The transmitting and receiving devices in a data transmission system must operate in step with each other to allow communication between them. Determining and maintaining the correct timing for transmitting and receiving information is known as *synchronization*. Special equipment must be provided to accomplish synchronization.

Most data transmissions are serial in nature; that is, one bit is sent behind the other. Both control bits and message data are contained in the same message stream. There are two main forms of transmission: asynchronous and synchronous.

Asynchronous transmission is often referred to as start/stop transmission because additional start and stop bits are transmitted with each character to identify the beginning and the end of the group of data bits. With asynchronous transmission, characters can be sent at irregular intervals. Synchronization is accomplished on a per-character basis, and synchronization of the terminals is reestablished as each character is received.

In *synchronous transmission*, characters are sent in a continuous stream without framing bits between characters. The sending and receiving devices operate continuously at essentially the same frequency and are synchronized, or kept in step, by electronic clocking instruments. See Figure 7.16.

The two types of transmission may be differentiated by the fact that in asynchronous transmission each character is transmitted as an independent entity with start and stop bits to indicate to the receiving device that the character is beginning and ending. In synchronous transmission, whole blocks of data are transmitted in units.

Asynchronous transmission is useful when transmission is irregular. It is less expensive than synchronous transmission because it requires less sophisticated circuitry. Synchronous transmission makes more efficient use of the media; it permits higher data transmission speeds because of the elimination of the start and stop bits.

Predominant Information Codes

Codes for representing information vary in two respects: the number of bits used to define a character and the arrangement of bit patterns in each particular character.

There are many transmission codes in use today. Some of the more commonly used codes are the International Telegraph Alphabet (ITA). No. 2, the American Standard Code for Information Interchange (ASCII), and the Extended Binary Coded Decimal Interchange Code (EBCDIC).

International Telegraph Alphabet No. 2

ITA No. 2 code replaces, and has many of the same characteristics as, the Baudot code. The principal use of ITA No. 2 is in international telex transmission. A description of the Baudot code will illustrate the principles of code construction and provide a basis for understanding more extensive coding systems

The *Baudot code* is a 5-bit code used by older teletype machines. (Figure 7.11 illustrates a 5-bit code.) The code is sequential; that is, a particular *control character* defines the subsequent series of characters until a new control character appears. There are 32 different combinations in any 5-bit code ($2^5 = 32$). The two control characters, letters-shift (LTRS) and figure-shift (FIGS), change the meanings of subsequent characters in much the same way as the shift lock on a typewriter does. Using letters and shift characters increases the number of available code configurations to 62 (64 − 2 shift characters). However, since 3 codes plus blank are the same in either shift, there are really only 58 different characters.

The Baudot code had several limitations:

1. It was very limited in punctuation and special character codes.
2. It had no parity bit or inherent method for validating transmission accuracy.
3. The sequential nature of the code meant that if a control character was missing, an entire portion of the message would be unintelligible.

Figure 7.17
The ASCII Code Configuration of Alphanumeric Characters

	Bit 7654321		Bit 7654321		Bit 7654321		Bit 7654321
A	1000001	**K**	1001101	**U**	1010101	**1**	1000110
B	0100001	**L**	0011011	**V**	0110101	**2**	0100110
C	1100001	**M**	1011001	**W**	1110101	**3**	1100110
D	0010001	**N**	0111001	**X**	0001101	**4**	0010110
E	1010001	**O**	1111001	**Y**	1001101	**5**	1010110
F	0110001	**P**	0000101	**Z**	0101101	**6**	0110110
G	1110001	**Q**	1000101			**7**	1110110
H	0001001	**R**	0100101			**8**	0001110
I	1001001	**S**	1100101			**9**	1001110
J	0101001	**T**	0010101			**0**	0000110

The early Baudot code has been replaced by ITA No. 2, an asynchronous code commonly—but incorrectly—referred to as the Baudot code.

Standardization of Coding Systems

During the 1960s a great many codes were developed, presenting the systems designer with a wide array of choices. It soon became apparent that some standardization would be desirable so that different computers and terminals could communicate. As a result, the American National Standards Institute developed the American Standard Code for Information Interchange (ASCII). This code has been adopted by the government and the military as their standard.

ASCII

ASCII is a 7-bit code, plus one bit for parity per character. The 7 bits provide 128 characters ($2^7 = 128$). Transmission may be either asynchronous or synchronous. ASCII is used more extensively than any other code in the United States because it is a standard in the communications industry. Its alphanumeric codes are shown in Figure 7.17.

EBCDIC

EBCDIC is IBM's System 360/370 code. It is an 8-bit code with 256 characters ($2^8 = 256$). It is generally transmitted in synchronous systems and has no provision for parity bits, although some users have modified it to provide for them. Because of the many possible ways to make the modification, users often ended up with an incompatible interface even though they were using the same basic coding system. Its alphanumeric codes are shown in Figure 7.18.

Figure 7.18
The EBCDIC Code Configuration of Alphanumeric Characters

Bit 87654321		Bit 87654321		Bit 87654321		Bit 87654321	
a	10000001	u	00100101	M	00101011	1	10001111
b	01000001	v	10100101	N	10101011	2	01001111
c	11000001	w	01100101	O	01101011	3	11001111
d	00100001	x	11100101	P	11101011	4	00101111
e	10100001	y	00010101	Q	00011011	5	10101111
f	01100001	z	10010101	R	10011011	6	01101111
g	11100001	A	10000011	S	01000111	7	11101111
h	00010001	B	01000011	T	11000111	8	00011111
i	10010001	C	11000011	U	00100111	9	10011111
j	10001001	D	00100011	V	10100111	0	00001111
k	01001001	E	10100011	W	01100111		
l	11001001	F	01100011	X	11100111		
m	00101001	G	11100011	Y	00010111		
n	10101001	H	00010011	Z	10010111		
o	01101001	I	10010011				
p	11101001	J	10001011				
q	00011001	K	01001011				
r	10011001	L	11001011				
s	01000101						
t	11000101						

Specialized Data Transmission Techniques

Chapter 6 discussed amplitude modulation and frequency modulation. In addition to these basic types of modulation, there are two specialized types: phase modulation and pulse code modulation.

Phase Modulation

As used in data transmission, the word *phase* refers to the relative timing of an alternating signal. Two signals of the same frequency differ in phase if one signal is behind the other by any amount that is not an exact multiple of the frequency. A sine wave starts at what is known as the *baseline,* rises to its peak at its 90-degree point, returns to the baseline at its 180-degree point, continues to its lowest point in the negative direction at its 270-degree point, and returns to the baseline at its 360-degree point. Thus, it completes one cycle. It then starts a new cycle and repeats the process indefinitely unless it is interrupted. (See Chapter 5, Figure 5.22 for an illustration of sine waves.)

In *phase modulation,* the phase of the signal is shifted to respond to the pattern of the bits being transmitted. In shifting the phase, the sine wave rises to its peak, then returns to the baseline at its 180-degree point. Then, instead of continuing in the negative direction, it starts upward toward another peak; that is, toward a new 90-degree point. Where no shift occurs, the signal is represented by alternating bits; that is, 0,1,0,1,0,1. To shift the phase, the signal would

Figure 7.19
Phase Modulation

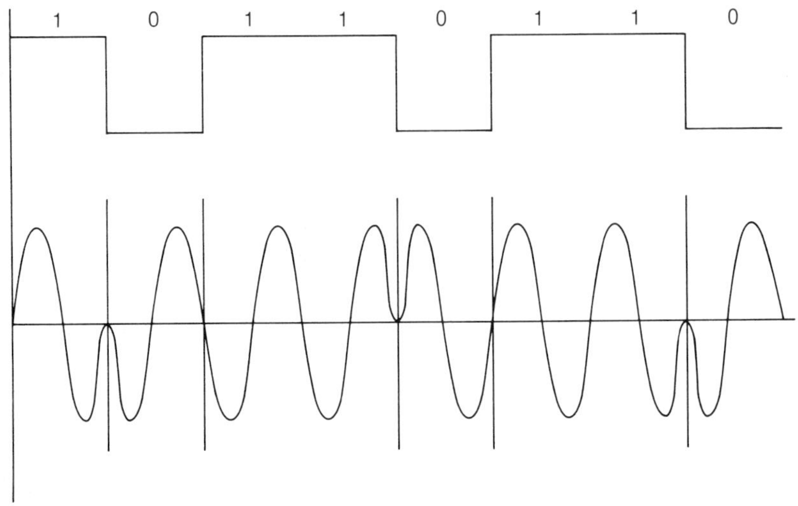

be represented by either two 1's or two 0's adjacent to each other; i.e., 1,0,1,1,0,1,1,0 (Figure 7.19).

When very high transmission speeds are required, phase modulation is preferred over other systems of modulation. Phase modulation is less affected by extraneous noise and is less error prone than frequency modulation.

Two-phase modems employ dibits, groups of two binary bits. This permits twice as many bits to be sent over the line in the same bandwidth in the same amount of time, thus increasing its speed. Although phase modulation is more costly than amplitude or frequency modulation, its speed and efficiency make it very valuable for data transmission.

Pulse Code Modulation

A frequently used technique in data transmission is *pulse code modulation (PCM)*. PCM converts analog signals into a series of coded digital pulses for transmission. At the receiving end the pulses are converted back into analog signals (Figure 7.20). PCM uses a fast sampling technique in which binary signals entered into the transmission system are representative of the entire signal. Many signals are sampled simultaneously in such a way that there is no interference among samples of the transmitted signals. Pulse code modulation results in a highly reliable, noise-free digital signal. This technique offers a very cost-effective way to increase the capacity of transmission lines and to improve the efficiency of data transmission.

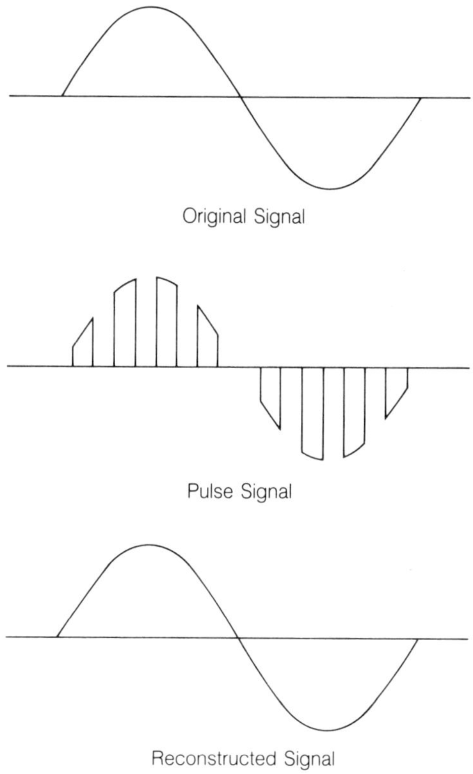

Figure 7.20
Pulse Code Modulation

Line Configurations

This chapter has explored the many facets of a data communication system. Regardless of the physical components of the system and the electronic techniques employed, the terminals and computer systems must be arranged in some type of line configuration. There are two principal types of line configurations: point-to-point lines and multidrop or multipoint lines.

Point-to-Point Lines

The simplest network consists of one circuit between two points and is called *point to point*. As the name implies, point-to-point lines directly connect two points in a data communication network (Figure 7.21). These lines are relatively expensive because each terminal uses a different line into the computer system.

Multidrop or Multipoint Lines

If additional terminals are added to the point-to-point line, it becomes a *multipoint* network. A multipoint or multidrop line has more than one terminal connected to the computer system. However, only one terminal can transmit to the computer at a time. See Figure 7.22.

Figure 7.21
Point-to-Point Line Configuration

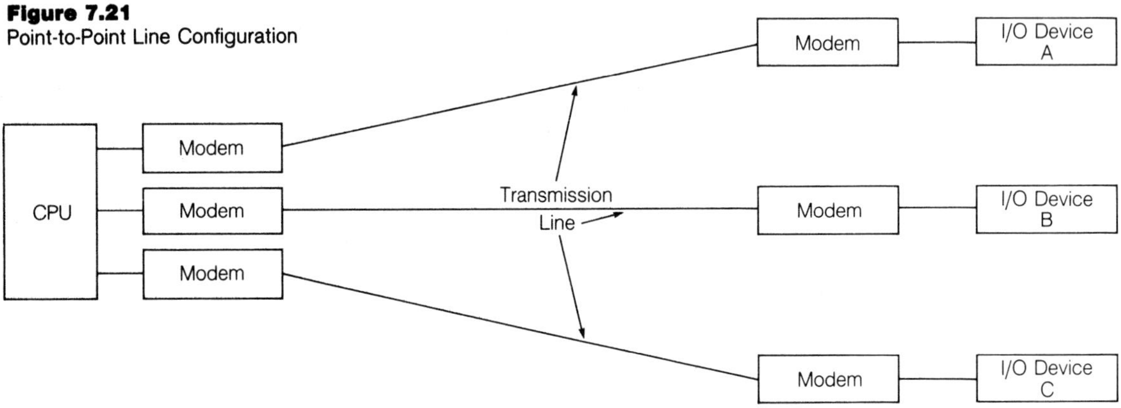

The number of terminals that can be served by one line depends on how much they are used. Multipoint operation is substantially less expensive than point-to-point operation because the per-line terminal cost drops appreciably when the line is used by a number of terminals.

Summary

Data communication systems are designed to transmit data from one location to another—usually from one computer or terminal to another. The data is transmitted in coded form over electrical transmission facilities.

Figure 7.22
Multipoint Line Configuration

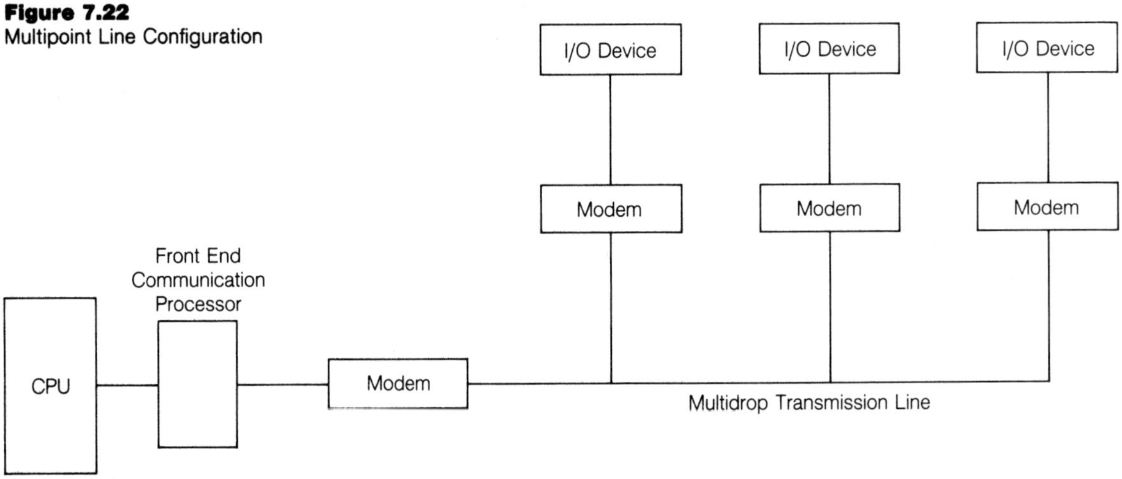

The three essential components of a data communication system are the source (remote terminals), the medium (transmission links), and the sink (the computer).

Five broad categories of terminals are: typewriter or keyboard terminals, video terminals, transaction terminals, intelligent terminals, and specialized terminals—audio response units and pushbutton telephones.

Telecommunication lines were originally built to carry voice communications consisting of analog signals. Since terminal devices transmit digital signals, an interface device is required to convert the data to analog form so that it can be transmitted over telecommunication lines. Where the transmission line carries digital signals, no modem is required. Two types of modems are: hard-wired modems, which are permanently connected; and acoustic couplers, which are portable.

Data transmission controls specify the rules to be followed in transmitting data (protocols), detect errors in data transmission (parity checking), and ensure system compatibility. Two levels of parity error detection are character checking and block or longitudinal redundancy checking.

Coding and sequencing should be sent properly so that the devices receiving the data signals will be able to interpret them as intended. The transmitting and receiving devices must operate in step with each other, or in synchronization. Two forms of transmission are asynchronous and synchronous. In asynchronous transmission, each character is transmitted as an independent entity with start and stop bits to indicate to the receiving device that the character is beginning and ending. In synchronous transmission, whole blocks of data are transmitted in units. Synchronous transmission makes more efficient use of the media; it permits higher data transmission speeds because of the elimination of the start and stop bits.

The ITA No. 2 code is commonly—but incorrectly—referred to as the Baudot code. This code is not used for data transmission; however, it provides a basis for understanding more extensive coding systems. The predominant transmission code today is ASCII; it is a standard in the communications industry.

Two specialized types of modulation are phase modulation and pulse code modulation. In phase modulation, the phase of the signal is shifted to respond to the pattern of the bits being transmitted. When very high transmission speeds are required, phase modulation is preferable. It is less affected by extraneous noise and is less error-prone than other systems.

Pulse code modulation uses a fast sampling technique whereby the binary signals entered into the transmission system are representative of the entire signal. It results in a highly reliable, noise-free digital signal. This technique offers a very cost-effective way to increase the capacity of transmission lines and to improve the efficiency of data transmission.

The two principal types of line configurations are point-to-point lines and multidrop or multipoint lines. Point-to-point lines are relatively expensive because each terminal uses a different line into the computer. Multipoint operation is less expensive than point-to-point operation because the per-line terminal cost drops appreciably when the line is used by a number of terminals.

Review Questions

1. Name and identify the three essential components of a data communication system.
2. Name several different categories of terminals and describe them briefly.
3. What are some of the factors to be considered in terminal selection?
4. Name and describe briefly the two principal types of modems.
5. What are protocols and what is their purpose?
6. What is the purpose of parity checking?
7. Briefly describe the two forms of synchronization.
8. What is the predominant transmission code today? Why is this code predominant?
9. What is the purpose of phase modulation, and how is it accomplished?
10. What is the purpose of pulse code modulation, and how is it accomplished?
11. Name and briefly describe two types of line configurations.

References and Bibliography

Asten, K. J. *Data Communications for Business Information Systems.* New York: The Macmillan Company, 1973.

Data Communications. *Executive Guide to Data Communications,* 5th Volume. New York: McGraw-Hill Publications Company, no date.

Davenport, William P. *Modern Data Communications.* Rochelle Park, N.J.: Hayden Book Company, Inc., 1971.

FitzGerald, Jerry, and Tom S. Eason. *Fundamentals of Data Communications.* New York: John Wiley & Sons, Inc., 1978.

Gillman, R. J., and D. A. Steedman. "CCITT Recommendations: The Use of Teletext Protocols in Store-and-Forward Systems." *Telephony*, May 2, 1983, 79–82.

Godin, Roger J. "Voice Input-Output." *Electronics*, April 21, 1983, 126–27.

Hindin, Harvey J. "System Integration: Local-Net Standardization Gains." *Electronics*, March 24, 1983, 98–99.

Martin, James T. *Telecommunications and the Computer*, 2d ed. Englewood Cliffs, N.J.: Prentice-Hall, Inc., 1977.

NCC Publications. *Handbook of Data Communications.* Manchester, England: The UK Post Office, 1975.

Sherman, Kenneth. *Data Communications*. Reston, Va.: Reston Publishing Company, Inc., 1981.

Techo, Robert. *Data Communications*. New York: Plenum Press, 1980.

8 Telecommunication Services

Our postindustrial society is often described as a service economy; that is, more workers are employed in providing services than in producing goods. Webster defines *services* as "useful labor that does not produce a tangible commodity." Many persons earn their livelihood by performing services for others—lawyers, delivery people, educators, barbers, ministers, bank tellers, nurses, custodians, brokers, and others.

A major objective of all organizations—whether business, government, or nonprofit—is maintaining or improving services while at the same time containing or reducing costs. Faced with ever-escalating labor costs, organizations are constantly seeking methods to control these costs. One such method is to automate part of the services that the operation requires. For example, using cash registers that automatically compute sales taxes and the amount of change to be returned to the customer simplifies the job of the sales clerk.

Another technique that has been widely accepted is getting the user of the service to do part of the job, a process often described as "externalizing labor costs." Supermarkets implemented this concept some time ago by providing shopping carts to customers, enabling them to pick out their own merchandise. By collecting their own items, customers perform a service previously provided by grocery clerks. Self-service is based on the idea that if the merchant is relieved of providing some services, operating costs will decrease and products or services can be sold for less. Many retail establishments have moved toward self-service by providing fewer clerks and expecting customers to locate and select merchandise for themselves.

Some companies have established service centers where product owners bring products needing repair, thereby reducing or eliminating repair service calls. Doctors, too, have embraced this concept. By virtually discontinuing house calls, doctors now require patients to travel to the service.

Automotive service stations have long provided equipment for customers to fill their own tires with air. Now, however, most gas stations are self-service; that is, customers pump their own gas. In fact, a person looking for a "full-service" station may have a difficult time finding one.

The communications industry has met the challenge of rapidly increasing labor costs and communications demand with new products that make it possible for customers to direct-dial many calls that previously required the assistance of an operator. The first long distance direct-dial service was inaugurated in 1956 by AT&T. Today it is possible for customers to direct-dial calls not only to anywhere in this country but also to most overseas locations. Direct-dial rates are substantially lower than rates for operator-assisted calls. Thus, the customer performs part of the service but pays less for the call.

Telephone subscribers can also save money by performing part or all of their own installation and maintenance work. The use of modular components has made it possible for customers to connect new or replacement telephone equipment to their systems. Equipment may be obtained from a customer service center or phone store. This reduces or eliminates the need for telephone installers or repairpersons to travel to customers' premises.

Telephone Services

Each of the four subsets of telecommunications—voice, data, message, and image—has its own complement of service offerings. Although some of these categories overlap, we will discuss them separately in order to help put them into perspective.

Local Exchange Services

The telephone office (or offices) that provides service for a specific geographical area is an *exchange*. An exchange may consist of one or several telephone offices and includes the physical plant and equipment necessary to provide communication services to the area.

Local exchange service is public telephone service to points within the designated local service area (exchange area) for a telephone. The local service area for a central office is usually defined in the telephone directory; typically it includes customers served by other nearby central offices. Service is charged at either a fixed monthly rate or by the amount of usage; calls are not billed individually.

An *exchange area* or *local service area* is a geographical area that has a single uniform set of charges for telephone service. An exchange area may be served by a number of central offices. A call

Figure 8.1
A Telephone Exchange Map

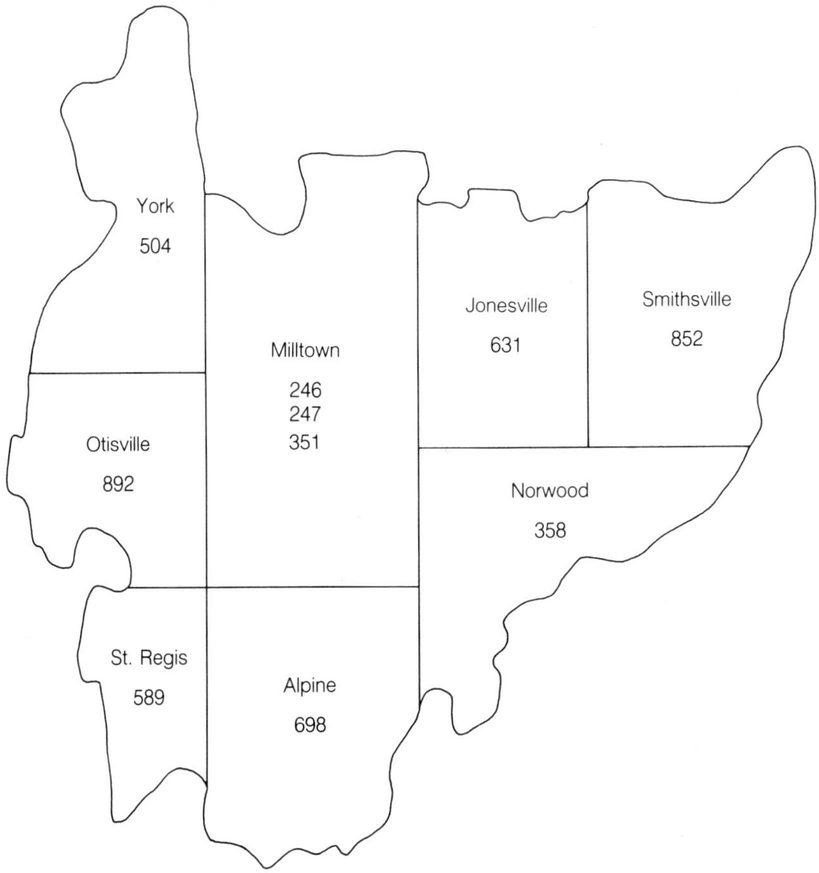

between any two points within an exchange area is a *local call.* Figure 8.1 illustrates a telephone exchange map of the areas that can be called locally from a given telephone location.

Rates for local exchange service are assessed in either of two ways:

1. *flat rate service,* wherein the user is entitled to an unlimited number of telephone calls within a specified local service area for a fixed monthly rate
2. *measured rate service,* wherein a charge is made in accordance with a measured amount of usage

Historically, *message units* were used as the unit of measurement in charging for local calls. Criteria used in determining the number of message units per call were the length of the call and the distance involved. For example, four message units might mean four separate calls of short duration to nearby locations. Or it might mean one or two calls of either longer duration or to a more distant location, or a combination of both time and distance.

A newer concept in measuring usage is *measured local service*. Under this concept, criteria used in measuring usage are the number of telephone calls, the duration of each call, and the distance involved. Thus, a new dimension is added to the measurement process by including the number of telephone calls as well as the duration of the calls and the distance involved.

Individual service is one telephone line that serves one subscriber. *Party-line service* is one telephone line serving more than one subscriber. Individual service is available to both residential and business customers. Party-line service is available to residential customers but seldom available to business customers.

Extended Area Service (EAS) may be purchased in many localities. This service permits a subscriber to make calls to a designated area beyond the local exchange area and be charged local exchange rates instead of toll rates. Subscribers in a metropolitan area who make a great many calls to a suburban locality beyond the subscriber's local exchange area can save money by purchasing extended area service rather than paying toll rates on the calls.

Long Distance Services

Long distance services are also called intercity services or toll services. They include any calls beyond the local service area and are charged under a tariff separate from local exchange tariffs.

Direct-Distance Dialing (DDD) This telephone service permits users to dial telephones beyond their local service area directly.

Operator-Assisted Calls These calls require an operator at a switchboard to complete the call. They cost substantially more than dialing direct. Operator-assisted calls can be classified as:

- ☐ person-to-person calls, in which the caller wishes to reach a particular person or extension number
- ☐ collect calls, in which the person or firm being called agrees to pay the charges for the call
- ☐ calling-card service (formerly known as credit-card service), in which callers who have a telephone calling card can have long distance calls charged to their regular monthly bill
- ☐ third-number calls, in which long distance calls are billed to an authorized third telephone number (a telephone number different from that of the calling or called telephone)

Although we have classified calling-card service as operator-assisted service, it should be noted that many pushbutton telephones now have a mechanized calling-card service in which the caller dials these calls without the assistance of an operator. Also, some public telephones are now equipped for handling calling-card calls.

Another service available on all types of operator-assisted long distance calls is *time and charges*. A caller can obtain this service by asking the operator for the length and cost of any call. There is an additional charge for this service.

Wide Area Telecommunications Service (WATS) The WATS system permits customers to make (outward WATS) or receive (800 service) long distance voice or DATAPHONE calls and have them billed at a bulk rate instead of at an individual call basis.

There are two types of WATS services available—intrastate and interstate. Intrastate WATS permits calling within a state. Interstate WATS permits calling within a specified band or bands extending outward from the originating state. It does not include the state in which the call originated. For interstate WATS, the United States is divided into a number of service areas, or *bands*, extending outward from, but not including, the customer's home state. Each service band subscribed to also includes the area or areas closer to the customer's home state. For example, if a customer living in band zero subscribed to service in band four, the customer would also be entitled to receive service in bands one, two, and three, but not in band zero.

800 Service Formerly known as inward WATS, this service permits one to call an office toll free from either intrastate or interstate locations, depending upon the type of service selected.

WATS is provided within selected areas by means of special private-access lines connected to the public telephone network. A single-access line permits either inward or outward service, but not both. In order to obtain both inward and outward services, two access lines are required.

Although WATS was formerly available in a number of configurations, it is now available only as measured service. Charges are based upon the number of subscribed service bands, amount of usage, and time classification in which usage occurs. For measuring usage charges, the hours of the day and days of the week are divided into rate periods, with lower rates in effect for evening and weekend periods. Thus, charges for calls of the same duration to the same service band vary depending upon the day of the week and the time of day the call takes place.

Enterprise Service This long distance feature permits an organization to have special listings in foreign city telephone directories.

Instead of listing the organization's local telephone number, the directory listing would read "Enterprise 1234." This service permits a caller in a foreign city to reach the desired number without incurring a toll charge. To call an Enterprise number, the caller dials the operator and asks for the Enterprise number. The operator translates this number into the local telephone number and places the call. The called party is billed for an operator-assisted call on a collect, toll basis.

Enterprise service is generally used when the volume of calls from foreign points is not sufficiently large to justify a toll-free number or when the geography of incoming calls does not match WATS line bands.

Private-line Service With this service, the customer has the exclusive use of a leased circuit between two specific points. The circuit is not connected with the public telephone network. Private lines may be used for any type of transmission—voice, data, teletypewriter, video, etc. (Private-line service should not be confused with individual service, a public network service in which one telephone line serves one subscriber.)

Private networks are a configuration of private lines and related switching facilities that are provided for the exclusive use of one customer.

Foreign Exchange Service (FX) This is service to a telephone exchange outside of the one in which the user is located. A leased line connects the subscriber's telephone to a central office in the foreign exchange area. The subscriber can be listed in the foreign directory. Being a two-way service, FX permits the subscriber to call any number in the foreign exchange area and people from the foreign exchange area to call the subscriber. This service allows users to avoid long distance charges to the foreign exchange.(For a more complete description of FX service, see Chapter 5.)

Specialized Telephone Services

Telephone subscribers are generally familiar with most of the traditional residential or business services because they have used them or at least are aware of their existence. There is another group of telephone services, however, that are perhaps not so well known—specialized services that are designed to meet a special need. They have one thing in common—people use them while away from their home or place of business. The two principal types of specialized services are coin telephone services and mobile telephone services.

Coin telephones These telephones fill a need that all of us have at one time or another—the need to make a call while away from our own telephone. There are two basic types of coin telephones: public coin telephones and semipublic coin telephones.

Public coin telephones provide service on the public network to persons away from their residence or place of business. They are installed in public areas such as airports, hotel lobbies, stores, and outdoor locations. The telephone numbers for public coin telephones are not listed in telephone directories.

Semipublic coin telephones are installed where there is a combination of general public and individual customer need for the service, such as in a gasoline station. With these telephones the subscriber receives a listing in the telephone directory and guarantees a specific monthly revenue from the telephone. The revenue is offset by the coins collected in the coin telephone. If the revenue from outgoing calls is less than the guaranteed amount, the customer is responsible for the difference.

A person who places a long distance call from a coin telephone elects either to pay for it immediately or to use a billing option such as calling-card service, third-number call service, or collect call service.

Most coin telephones now provide *dial-tone-first service,* which permits customers to reach the operator and to dial certain calls, such as directory assistance or 911, without depositing a coin (911 is a code used for emergency calls in many localities). The dial-tone-first feature gives some assurance to the user that the telephone is working before coins are deposited. It also allows callers to use the telephone more easily in emergency situations.

Mobile telephone services utilize radio transmission. These services include land mobile service, paging service, air/ground service, VHF maritime service, coastal harbor service, high-seas maritime radiotelephone service, and high-speed train service.

Land Mobile Telephone Service This service provides two-way voice communications, through mobile-equipped central offices, between mobile units and land telephones or between two mobile units.

Paging Service This personal signaling service notifies a user to contact the control dispatch point to receive a message. A person contacts the individual by dialing a special telephone number. This causes an audible tone to be emitted from a small radio receiver that the user carries. Many doctors and field service personnel use paging service.

Air/Ground Service Such service allows two-way telephone communication between aircraft in flight and parties on the public telephone network. The service is provided by radio base stations connected to control terminals and mobile service switchboards. A few airlines offer this service on some of their premium flights.

Marine Radio Telephone Services Ships at sea use this for two-way telephone service. There are three types of service, which differ basically in the distance range in which they operate:

1. VHF maritime service, which provides reliable communications in the very high frequency band up to 50 miles offshore
2. coastal harbor service, which provides communications up to 1,000 miles offshore
3. high-seas service, intended for ships engaged in high-seas operations and transoceanic passages

Telephone Support Services

Local telephone companies provide support services designed to enhance the utility of the company's service offerings. These services include business office services, community services, telephone directories, directory assistance, and intercept services.

Business Office Services The business office takes orders for new services or for changes in service, answers billing inquiries, receives bill payments, and completes the link from the customer to the rest of the telephone company's working forces. Service representatives coordinate these functions with customers.

Community Services Some telephone company services are of a civic nature; that is, they are provided primarily as a public service to the community. The program to make 911 the emergency reporting number is one example. Dialing 911 in many areas of the United States puts the caller in direct contact with the local police, fire department, or emergency medical service. In addition, many localities provide emergency reporting telephones connected directly to the emergency center. They are installed in public areas that are readily accessible. The cost of this service is borne by the providing locality; the user pays no charge for the call.

Telephone Directories The White Pages contain an alphabetical listing giving each subscriber's name, address, and telephone number. For an additional fee, users select special directory listings, such as "If no answer, call . . . ," "After 5 o'clock call . . . ," or boldface type. An additional service is the withholding of a customer's listing from the directory (nonlisted number). This service offers some measure of privacy, but because it involves special handling, telephone companies charge an additional fee.

The Yellow Pages contain an alphabetical listing of business subscribers by category of business. Subscribers can purchase advertisements in the Yellow Pages of the telephone directory.

In some areas, telephone directories also contain a section of Blue Pages that lists the numbers for frequently called organizations such as government offices, media, and emergency agencies.

Directory Assistance This service is designed to provide telephone numbers not included in the local telephone directory. Its use by customers unable or unwilling to look up local telephone numbers has grown to such proportions that many telephone companies now charge for the service. Directory assistance for foreign codes can be reached by dialing the area code and 555–1212. Although long distance directory assistance was traditionally a free service, in the present competitive market carriers that provide the service charge for it.

Intercept Service This service informs callers of any changes regarding the telephone number they have dialed. Calls to a disconnected number are automatically routed to an intercept operator. The operator informs the caller of the status of the number called and provides a new number if one is available.

In some locations automatic intercept systems (AIS) have been installed. These systems improve the processing of calls to nonworking numbers by automating service. Under AIS, calls in local offices are routed to a central intercept center. The files are searched for the called number, and a recorded announcement is connected to the customer's telephone. The announcement contains all the information available, including the number the customer dialed and the new number if one is available.

Telephone Service Features

The telephone services described thus far are provided by the local telephone company, with the user frequently providing some part of the service. The following services are inherent in the telephone equipment—the switching system or the telephone instrument. The equipment may be provided either by the telephone company or by an interconnect vendor.

Custom Calling Features The four features that are generally classified as custom calling features are three-way calling, speed calling, call forwarding, and call waiting. All of these features can be provided by electronic switching systems. In addition, speed calling can be provided by telephones with electronic components.

1. *Three-way calling* permits a customer to add a third party to an existing conversation for a telephone conference call. When the third party answers, a private, two-way conversation with that party can be held before bridging the connection for the three-way conference.
2. *Speed calling* (also called *automatic dialing*) permits a caller to reach certain frequently called numbers by using abbreviated telephone codes in place of the conventional telephone number.
3. *Call forwarding* permits a telephone to automatically forward calls to another telephone number. The equipment can generally be programmed so that the call is forwarded only after a predetermined number of rings. This feature can be used to transfer calls to either a local telephone or a telephone in a distant city. When a forwarded call is subject to a toll charge, the charge is billed to the forwarding telephone.
4. *Call waiting* permits a call to a busy telephone to be held while an audible tone notifies the called party that a call is waiting. The tone is audible only to the called party, who can decide whether to interrupt the existing conversation to find out who is calling.

There are many different service features available to subscribers with private internal telephone systems (PBXs or CBXs). Because these features vary from one system to another, in selecting a telephone system, the customer must evaluate the organization's needs carefully and match these requirements with the features. Not all features are available from all telephone equipment suppliers, but those listed here are available from most vendors.

1. *Automatic call back* permits the caller to "instruct" a busy station to call back as soon as the busy station is free. The instruction is given by dialing an extra digit that tells the computer to re-establish the connection when both telephones are available.
2. *Line privacy* is a station control feature for telephone lines that require special privacy. The feature is useful for data processing lines where the data could be contaminated or destroyed by outside interference. To obtain line privacy, the user activates a control key that excludes transmission interference by another person or electronic device.
3. *Lockout* is a station control feature that ensures the confidentiality of a call. Many newer telephone systems permit a switchboard attendant to break into a call in progress; the lockout feature allows the user to "lock out" interruptions until the call is completed.

4. *Automatic route selection (ARS),* sometimes called *least-cost routing,* permits the automatic selection of the most efficient routing of a call originating in a corporate network; e.g., tie line (first choice), WATS line (second choice), and direct-distance dialing (third choice). Some users can be denied access to the latter category, so that the call is delayed until one of the less costly routes becomes available.
5. *Trunk prioritization* enables a customer to use its WATS lines or other specialized facilities to their fullest by stacking up the calls of those users having lower priority. When desired lines are all occupied, this option records the number dialed and makes the connection when a line becomes available.
6. *Remote access* permits authorized personnel to place calls from other locations and be connected to a business PBX system. Users access the system by dialing a private telephone number plus a 3-digit security code. Then the authorized caller can place any type of call that could be placed from a PBX station.
7. A *message-waiting indicator* is a message light on a user's instrument that can be lighted when the receptionist or message center presses a button or transmits a dial code. A lighted indicator tells the user that a message is waiting. (This should not be confused with *call waiting,* wherein a caller is on the line trying to reach a telephone that is busy.)
8. *Identified ringing* provides distinctive ringing tones for different categories of calls. For example, internal calls, calls from a secretary, or calls from a given extension can be recognized by their unique ringing style. Thus, before answering the telephone the called person is given some indication of where the call originated.
9. The *call pick-up* feature enables a person receiving a telephone call to have access to the incoming call on any telephone station in the system by entering a code. Also, others in the called party's group can pick up calls to take messages. This feature enables a telephone system to operate effectively without having all lines appear on each telephone station.
10. The *individual call transfer* feature enables a telephone system user to transfer a call to another station without going through an operator, thus saving time for both users and operators.
11. *Paging access* allows an authorized station user to have direct access to paging equipment. The feature is activated either through dialing a special code or through pushing a paging button on the instrument.
12. *Recorded telephone dictation* equipment permits the user to be connected with central dictation facilities. It may also permit the user to access central dictation facilities directly from the public

Figure 8.2
Automated Repair Service Bureau

(Reproduced with permission of AT&T)

telephone network. This feature is useful for persons who are out of town or working at home.

13. A *local maintenance* feature enables a special console to be used to reprogram the telephone system to change system features or stations. Some PBX equipment has *remote maintenance* capability, through which a service person can dial the telephone system to gain access to the central processor of the system to test or modify system features or programs. (See Figure 8.2.)
14. *Electronic telephones.* Some telephone sets contain electronic components or microprocessors that enable them to perform a variety of special services. These telephones are described as "smart" or "intelligent" because of their enhanced capabilities. They can be designed to perform any or all of a wide variety of telephone services such as speed dialing, last number redial, hands-free operation, automatic answering, call timing, and many others.

Controlling Telephone Usage

Additionally, service features are available to control telephone usage. *Automatic Identification of Outward Dialing (AIOD),* also called *Station Message Detail Recording (SMDR),* are services that record call details. AIOD is the name of the recording system provided by interconnect vendors; SMDR is the Bell System's product. This equipment consists of a tape recorder with minicomputer features; it is activated on long distance calls made from a PBX system extension. Details of the call (date and time, originating extension number, called number, start and end times of conversation, type of line used) are recorded on tape. The tape is processed periodically, either in-house or by a service bureau, and a report summarizing the information obtained. This report is compared with the telephone company's billing so that the cost of each call can be allocated to the extension making the call.

Some systems do not record the calling station automatically, but require the caller to dial an identification code before the desired long distance number. Caller identification codes can also be used to restrict usage by unauthorized callers or to restrict usage to local calls. A complete listing of charges incurred permits supervision of telephone costs, which, in turn, can do much toward reducing an organization's telephone expenses.

Another way of controlling telephone usage costs is through service restrictions. There are two types of restrictions that are frequently employed: class of service restriction and originating restriction.

A *class of service restriction* feature limits use of a telephone station to certain types of calls. For example, a telephone station might be limited to only internal calls. With this feature, each telephone in a system can be provided with different levels of access to outside lines and telephone services. A telephone station might be restricted to accessing only certain cities or area codes. Or it might be restricted from using special long distance services such as WATS or FX lines.

An *originating restriction* feature restricts the telephone station from being used to place outgoing telephone calls. The station can receive calls in the usual manner. This restriction is often placed on telephones in conference rooms or unoccupied offices. The restriction is applied automatically; a person attempting to place a call by dialing "9" to obtain an outside line receives a busy tone. Some telephone systems restrict outgoing calls by requiring them to be handled by an operator.

The restrictions are customized to serve the needs of the organization. These features pay for themselves by controlling telephone costs.

Written Message Systems

The fastest, most efficient, and often the least expensive method of communicating is by telephone. The telephone provides two-way conversation, allowing exchanges of ideas and information. When a written record of the message is required for reference or legal purposes, the telephone must be supplemented by a written document.

The business letter is undoubtedly the most widely used form of written message communication. However, its production and delivery is a costly, time-consuming process. A letter must be composed, typed, placed in an envelope, stamped, delivered to a mailbox, picked up and taken to a post office, sorted again, and physically delivered to its destination. At its destination it frequently requires additional sorting and delivery to the proper department and person.

The many operations involved in the manual delivery of written communications make it a slow process. Further, since these operations are highly labor-intensive, they are expensive.

Electronic Mail

Electronic methods of sending messages offer a viable alternative to traditional message services. *Electronic mail* refers to the delivery of mail, at least in part, by electronic means. Messages can flow faster when they are distributed electronically because they do not have to be physically handled as much.

Messages can be sent electronically over any telecommunication facility including telephone lines, microwave radio, waveguides, coaxial cable, satellites, and fiber optics. Messages are transmitted and received in the form of electronic signals that are translated into readable messages by the receiving device (facsimile machine, teletypewriter, computer printer).

The major categories of electronic mail systems are facsimile systems, teletypewriter systems (Telex and TWX), communicating word processors, carrier-based message systems, and private computer-based message systems.

Facsimile Systems One of the most dependable ways of transmitting information is by the time-proven technology of facsimile, used in offices for over fifty years. Facsimile, abbreviated FAX, is a system for transmitting a copy of a document to a distant point over telephone lines. Facsimile machines are also known as *telecopiers.* Connected to each other via office telephones, these machines can transmit and receive any form of documents in minutes—or even seconds. The process can be compared to putting a copy of a document into a copier at one location and having it come out at another location.

To use a facsimile machine, the operator places the original document on a transmitting tray, establishes phone contact by coupling the telephone receiver to the sending device, and presses the appropriate key to start transmission. The transmitting device scans the

Figure 8.3
Canon Facsimile Copier

(Courtesy of Canon, U.S.A., Inc.)

entire page of the document line by line using a light source (lens, laser, or fiber optic), converts the document into digital signals, and transmits the digital data to the receiving machine. There it is reconverted, and the document is reproduced in its original form.

In addition to transmitting typewritten or printed text, facsimile machines can send handwritten text, photographs, and graphic materials. Since facsimile machines accept original documents, no rekeying of text is necessary and accuracy of output is assured. See Figure 8.3.

Facsimile units are available in low-speed (4–6 minutes per page), medium-speed (1–4 minutes per page), and high-speed (seconds per page) models. The higher speed models are more costly, but because of their shorter transmission time they often save money in the long run. The lower speed units produce a clearer copy, making them particularly useful for transmission of graphic materials.

Aside from business applications, a large part of facsimile usage is in telephotography, the transmission of news pictures. The U.S. National Weather Service faxes weather maps to forecast centers all over the country, which use them to prepare forecasts for various media, airports, and ocean-going vessels.

Teletypewriter Services Teletypewriters are electronically controlled typewriters that send and receive messages over communi-

Figure 8.4
A Teletypewriter

(Reproduced with permission of AT&T)

cation lines (Figure 8.4). Teletypewriter machines can be obtained from Western Union or other vendors. They can be connected by public or private telephone lines or by Western Union facilities. In addition to communicating with another subscriber, teletypewriters may be used to send Mailgrams, telegrams, cablegrams, and money orders. A variety of specialty services, such as news and financial information services, are also available on a dial-up, subscription basis.

Western Union operates both Telex and TWX local and long distance services. Both are dial-up services wherein a subscriber can dial any other subscriber and transmit messages. A Western Union directory called *Infomaster* provides the names, locations, and identification of all subscribers.

Telex/TWX Combined Service (TCS) enables Telex and TWX customers to communicate with each other. Since Telex transmits on a 5-channel tape at 66 words a minute, and TWX transmits on an 8-channel tape at 100 words a minute, the systems are basically incompatible. TCS service translates the form of signal carried on the transmission line so that these two types of message devices can communicate with each other.

Communicating Word Processors Some word processors are equipped with electronic components that enable them to send documents from one location to another over telephone lines or other telecommunication channels. When two communicating word processors are connected to each other, they are said to be *online* and can send and receive messages.

To input a message, the user keyboards onto a magnetic medium. When the text is ready for transmission, the operator dials the phone number of the receiving machine, places the telephone receiver in the word processor's coupler, and the message is transmitted over telecommunication facilities. At the receiving terminal, the message can either be stored on a magnetic medium or printed as hard copy. Messages can be sent at any time of day or night to either attended or unattended terminals. This permits low-priority messages to be sent during reduced telephone rate periods.

Because of their two-way communications capability, communicating word processors can serve as message terminals for electronic mail service. They can be used as substitutes for conventional message equipment such as teletypewriters. Since word processors also perform other office functions, the inclusion of communication capabilities enables them to serve dual purposes, eliminating the need for expensive teletype machines.

Carrier-Based Message Systems These systems offer electronic mail service to the general public. Service features differ from carrier to carrier. Some services are fully electronic end to end; others supplement electronic transmission by U.S. Postal Service or messenger delivery. Carrier-based systems include Western Union's telegrams, Mailgrams, and EasyLink INSTANT MAIL; MCI Mail; and Federal Express's ZapMail.

Western Union offers two basic types of telegraph services: the *telegram*, which is a priority service, and the *overnight telegram*, which is a service with following morning delivery. Delivery of either type of telegram is by telephone or teletypewriter unless the sender specifies message delivery. Physical delivery is available only in certain locations specified in the Western Union directory. Telegrams carry liability for mistakes or delays in delivery. Tariffs filed with the FCC set minimum delivery service standards for telegrams.

Mailgram is a fast communication service offered jointly by Western Union and the U.S. Postal Service. Mailgrams are transmitted electronically over Western Union's microwave network to the post office nearest the destination address. If filed by 7:00 p.m., they are delivered the next business day by regular mail carriers.

Messages may be entered directly into the Mailgram system by firms with teletypewriter service at a per message charge. Also, large volumes of messages can be entered directly into the Mailgram system from computer tapes prepared by either the customer or by Western Union.

Another Mailgram service is the *Stored Mailgram,* which permits the user to store frequently used information for later use. To access the information stored in the computer file, the customer dials a toll-free telephone number, identifies the account to be charged, and requests the computer to perform the desired service. The computer processes the messages and prepares them for Mailgram distribution.

A Western Union service that is available to customers who have electronic transmission equipment is EasyLink INSTANT MAIL. Message delivery is fully electronic if the recipient is also an EasyLink subscriber; otherwise, delivery is by U.S. Postal Service or courier.

Another service for users with their own electronic machines (computers, word processors, Telex machines, or electronic typewriters with memories) is MCI Mail. Users of this service must keyboard the messages themselves. Like Easylink INSTANT MAIL, message delivery is fully electronic if the recipient is also an MCI subscriber; otherwise, delivery is by U.S. Postal Service or courier.

Federal Express's ZapMail does not require the sender or the recipient to have electronic equipment. Documents are picked up by courier, transmitted via Federal Express facsimile machines, and delivered by courier to their destinations—all within a few hours.

Message switching is the process of routing messages among several locations using either circuit switching, as in a dial telephone system, or store-and-forward techniques in a computer. Message switching services may be obtained either by purchasing or leasing message-switching equipment (private message switching) or by using the services provided by a message switching vendor. Western Union, GTE, Telenet Corporation, and other common carriers provide public networks for message-switching services.

International communications depend heavily upon written-record services. These services include cablegrams (international telegrams), Telex and TWX services, and leased-wire services. Common carriers offering international written-message services include International Telephone and Telegraph (ITT) Worldcom, RCA Globecom, and Western Union International.

Private Computer-Based Message Systems Many general purpose computers can store messages and forward them between terminals having access to the CPU of the computer.

To send a message, the user establishes connection between the terminal and the computer, then keys in a user code, or password, assigned to each person authorized to use the system. The user then transmits the address of the terminal for which the message is intended, followed by the message itself. This is done by keyboarding the message, editing it on the terminal screen, and sending the message to the memory of the computer, where it can be held until the person to whom it is sent is ready to receive it. Messages are sent over telephone lines, using either dial-up or direct-line service.

To notify the addressed online terminal that a message is waiting to be delivered, the computer flashes a signal on the terminal screen of the intended recipient. The person to whom the message is being sent accesses the computer, enters the proper password, and the message is transmitted onto the screen. There it can be read directly and/or printed as hard copy if a record of the message is desired. A similar service, computerized voice answering, is available from several vendors including IBM, ROLM, and Wang.

To use the message-waiting feature, the message destination terminal must be online with the computer. If the destination terminal is not online, it will not be able to receive the notification that a message is waiting. However, a terminal user can initiate periodic inquiries to determine whether or not the computer is holding a message. If there is a message, it would be received in the same way as it would by an online terminal.

The use of computer-based message systems can save corporate executives a considerable amount of time. James Martin, writing in *Telematic Society*, says:

> In the United States, only about 68 percent of long distance calls are completed and only about 70 percent of local calls are completed. . . . On the completed calls, the called party is reached only about 25 percent of the time. In other words, less than 25 percent of all calls attempted reach their desired party.[1]

The exasperating phenomenon of trying to reach a person by telephone is sometimes referred to as "telephone tag." When a caller calls and the called party is not available, a message is left. Upon receiving this message, the recipient calls back, only to find that now the original caller is not in, and so on. Since computer-based message systems hold messages until the intended recipient is ready to receive them, they free business executives and secretaries from much of this tiresome, time-consuming process.

Because they need timely information to expedite management decision making, organizations are seeking alternatives to the relatively slow delivery provided by the United States Postal Service. The almost instantaneous speed of electronic mail makes it a very attractive alternative.

A deterrent to the use of electronic mail systems is their initial cost. Although the cost of transmitting messages electronically is low, the costs of procuring the required equipment can be substantial.

1. James Martin, *Telematic Society* (Englewood Cliffs, N.J.: Prentice-Hall, 1981), p. 90.

Thus, present use of this service is limited to those firms with sufficient volume to justify equipment costs.

Another drawback is that there is essentially no way to know who has the necessary transmission and receiving equipment. Except in the case of Western Union's Infomaster, there are no available directories of equipped stations.

In spite of these limitations, the future prospects for electronic mail are bright. As costs of office supplies increase and office personnel and postal workers receive higher salaries, the costs of physical mail delivery can be expected to rise. Costs of electronic mail, on the other hand, can be expected to drop sharply as a result of continued development in microprocessor, satellite, laser, and fiber optic technologies.

Data Services

Two basic services provided by common carriers enable data to be transported from one location to another electronically. They are digital to analog signal conversion and provision of digital facilities.

Digital to Analog Signal Conversion

Conversion of digital signals to analog signals enables data to be carried over existing analog transmission facilities. The conversion is accomplished by modems at both the originating and receiving ends. Modems can be obtained from a number of suppliers and are available in a wide variety of transmission speeds. By using modems for signal conversion, customers can use either private lines or the public telephone network (including WATS and FX lines) for data communications.

Digital Data System (DDS)

A digital data system uses interconnected digital transmission facilities that form a synchronous network for data communications. Since signaling is digital, no modems are required. Digital data systems are provided on a private-line basis between large metropolitan areas. They are available from many carriers.

Image Services

Image services communicate exact images such as pictures, blueprints, graphic displays, faces on picturephones, and freeze-frame television across distances. Our classification of telecommunication services as voice, message, data, and image is not absolute; some overlapping exists. Some written message services could also be classified as image because they produce both a written message and an image. Facsimile services and CRT displays of data fit this description.

Video services can be classified into five main categories: commercial television, cable television, satellite television, freeze-frame television, and conference-circuit television.

Commercial Television

Commercial television is the television that traditionally has come to us in our homes. It originates at a central location and is distributed to television broadcasting stations via a telecommunication network. The transmission facilities required to carry one commercial television program are equivalent to 600 voice-grade telephone lines. Television stations broadcast their signals through the air from their radio-like antennas. The signals are received on antennas and images are produced on television sets.

Cable Television

Cable television signals are carried over coaxial cable distribution systems. The television signals are transported from a central source directly to receiving television sets. Cable television is a subscription service that requires a special line connecting the subscriber to the distributing point. There are many more channels available on cable television than on commercial television.

Satellite Television

Satellite television provides a picture signal that is transported directly from its source to an orbiting satellite. The satellite reflects the signal back to dish antennas connected to conventional television sets. Satellite transmission provides a greater number of channels and a wider variety of programming than either commercial television or cable television. Since the cost of a dish antenna is substantial, satellite television is used primarily by commercial establishments.

Freeze-Frame Television

Freeze-frame television is a technique for sending pictures over voice-grade telephone lines from one location to another. Pictures do not appear continuously; they are held for a period of time until the next picture is transmitted. Freeze-frame pictures are used as an adjunct to teleconferencing services.

Conference-Circuit Television (Videoconferencing)

Conference-circuit televison service permits a group of people in one room to see and hear people at a different location on a two-way basis. Use of this service saves money and time spent in traveling between various locations.

Videoconference transmission requires high-quality, wideband facilities, resulting in relatively high costs. These costs must be weighed against the costs of time and travel required to bring both groups together in one location.

Summary

Organizations are constantly seeking ways to control costs. One way to do this is to get the user to do part of the job. The communications industry has incorporated new technology into their products, enabling the customer to perform many functions previously done by telephone company personnel. These functions range from installation and maintenance to direct-dialing of long distance calls. Each of these self-performed services results in cost savings for both the user and the telephone company.

Each of the four subsets of telecommunications—voice, data, message, and image—has its own complement of service offerings.

Voice services fall into four main categories—local exchange services, long-distance services, specialized telephone services, and telephone support services. Local exchange service is public telephone service to points within the local service area for a telephone; it typically includes customers served by other nearby central offices. Extended area service, which permits a subscriber to make calls to a designated area beyond the local exchange area without paying toll rates, is available in many areas.

Long distance services (also called intercity or toll services) include any calls beyond the local service area. They may be obtained on either a direct-dial or operator-assisted basis. Business users with large volume may find it advantageous to use WATS, enterprise, private-line, or foreign exchange service.

Specialized telephone services are designed to be used by persons away from their home or place of business. The principal types of specialized telephone services are coin telephones and mobile telephones.

Telephone companies also provide a number of support services to enhance the utility of their service offerings. These services include business office services, community services (direct contact with police, fire, and other emergency services), telephone directory, directory assistance, and intercept service.

Modern telephone equipment is also capable of providing a variety of enhanced features. These include custom calling (three-way calling, speed calling, call forwarding, and call waiting), automatic call back, line privacy, lockout, automatic route selection (least-cost routing), trunk prioritization, automatic identification of outward dialed calls (AIOD), station message detail recording (SMDR), and service restrictions.

Electronic mail (written message service) is the delivery of mail, at least in part, by electronic means. Its principal advantage is speed. The major categories of electronic mail systems are facsimile; teletypewriter (Telex and TWX); communicating word processors; carrier-based systems (telegram, Mailgram, EasyLink INSTANT MAIL, MCI Mail, ZapMail); and private, computer-based message systems.

There are two basic data-transmission services provided by common carriers. They are: digital to analog signal conversion, which uses modems that enable data to be carried over analog facilities; and digital facilities, which are interconnected to form a synchronous network for data communications.

Image services communicate exact images such as pictures, blueprints, graphic displays, and freeze-frame television across distances. The principal types of video services are commercial television, cable television, satellite television, freeze-frame television, and conference-circuit television.

Review Questions

1. What is a telephone exchange area?
2. Describe extended area service. When would it be advantageous to subscribe to this service?
3. Name two types of savings effected by the use of direct-distance dialing.
4. What is the advantage of wide area telecommunications service (WATS)?
5. If a customer subscribed to WATS band 5, what other WATS bands would the customer be entitled to call?
6. What is the purpose of telephone support services? Name several types of support services.
7. Describe two ways in which telephone custom-calling features can be provided.

8. What is the importance of PBX remote testing capability?
9. What is electronic mail? What is its chief advantage?
10. Describe the digital data system (DDS).

References and Bibliography

Arntson, L. Joyce. *Word/Information Processing.* Boston: Kent Publishing Company, 1983.

Bergerud, Marly, and Jean Gonzalez. *Word/Information Processing Concepts.* New York: John Wiley & Sons, Inc., 1981.

Cortes-Vergara, M. C. "Telephone Conferences Offer Savings Through More Efficient Use of Time." *Communications News,* February 1983, 83.

Crump, Stuart. "Cellular Radio Ushers in Revolutionary Changes for Servicing Business User Needs." *Communications News,* February 1984, 65–67.

Elton, Martin C. J. *Teleconferencing: New Media for Business Meetings,* AMA Management Briefing. New York: American Management Association, 1982.

Fross, Alan. "Answers to Frequently Asked Questions About Store and Forward Voice Systems." *Communications News,* April 1983, 34–35.

Goeller, L. F. *Voice Communications in Business.* Geneva, Ill.: abc of the Telephone, 1982.

Goodman, Danny. "Car Telephones: Cellular Technology Promises More Channels." *Radio-Electronics,* February 1982, 41–44.

Mathai, Thomas. "Electronic Mail." *Telecommunications,* March 1983, 56–58.

Martin, James T. *Telecommunications and the Computer,* 2d ed. Englewood Cliffs, N.J.: Prentice-Hall, Inc., 1977.

———. *Telematic Society.* Englewood Cliffs, N.J.: Prentice-Hall, Inc., 1980, 90.

Nabs, Charles N. "Strategic Planning for Telecom Management in the 1980s." *Business Communications Review,* Vol. 13, No. 3 (May–June 1983), 12–18.

Posa, John G. "Radio Pagers Expand Horizons." *High Technology,* March 1983, 44–47.

Reynolds, George W. *Introduction to Business Telecommunications.* Columbus, Ohio: The Charles E. Merrill Publishing Company, 1984.

Robbins, A. "The Cellular Telephone Goes on Line." *Electronics,* September 22, 1983, 121–29.

Rosen, Arnold, Eileen Feretic Tunison, and Margaret Hilton Bahniuk. *Administrative Procedures for the Electronic Office.* New York: John Wiley & Sons, Inc., 1982.

Shanahan, Sharon L. "Facsimile Joins the Computer Age." *Infosystems,* September 1982, 126–31.

Strock, Robert D. "Electronic Mail—The Evolving Business Environment's Natural Selection." *Telephony,* May 5, 1982, 24–32.

Principles of Telecommunications Management

9

Historical Perspective

Telecommunications in business and industry started out as a switchboard service managed by the telephone operator. The operator could receive calls, place outgoing calls, and handle internal calls using private branch exchange (PBX) equipment. As telephone usage increased, organizations sought ways to improve the service and to control costs. The result was the development of the private automatic branch exchange (PABX), which allowed internal and outgoing calls to be placed directly by the user. The operator handled only incoming calls.

The next development in telephone technology was the Centrex system, which permitted callers to dial the desired telephone number directly. Since Centrex systems completed most calls automatically, the role of the operator was virtually eliminated.

The principal tasks to be performed in connection with the traditional telephone system were checking the monthly bill and ordering repairs and equipment changes. Someone had to be responsible for performing these tasks. In most offices, this was an office manager, secretary, or clerk.

The Telephone Company as Manager

Since the person responsible for the ongoing operation of an organization's telephone system rarely had specialized expertise in this area, the local telephone company, sometimes referred to as "telco," served both as provider of services and consultant. In other words, the telephone company "managed" the telecommunications function of most organizations.

When an organization's telecommunications needs changed, it called upon the local telephone company for assistance. The telephone company made a study of the organization's call volume (time of day, destinations called, and length of calls), evaluated the data, and made recommendations. Because the telephone company furnished this service, many organizations found it unnecessary to establish internal management of their telephone system or to employ a consultant. There were few options in equipment or services and no options in price; therefore, few decisions had to be made.

Telephone Management

Telephones are standard equipment in any office; businesses could not function without them. The role of telephone management is to provide good service to an organization and its employees. This includes ensuring proper usage and controlling costs.

Operating conditions have changed dramatically since federal regulators and the courts modified the view that telephone companies should be regulated monopolies and started to open the telecommunications industry to competition. Yet some organizations are still managing their telephone systems in the same way as they did years ago. This usually entails:

1. ordering telephone service for anyone requesting it
2. relying entirely on the supplier of the services for their management
3. failing to control telephone usage

Without clearly defined guidelines concerning telephone services, overprovision and misuse nearly always result.

Need for Telephone Management

Telephone costs represent a substantial and ever-increasing portion of a company's budget. In many organizations, they are exceeded only by costs of labor and office space. Before competition, telephone costs were considered an uncontrollable overhead expense. Now, however, there are many alternatives available for reducing telephone costs.

Rapid advances in technology, along with deregulation in the telephone industry, have resulted in many new types of equipment and services, a multitude of new suppliers, and aggressive price competition. A business has many options in obtaining telephone equipment and services:

1. It can rent equipment from AT&T Information Systems.
2. It can buy equipment from AT&T Information Systems or an interconnect vendor.

3. It can enter into a third-party lease agreement.
4. It can lease telephone circuits from the telephone company or from other carriers. In most cases, the telephone circuits can be leased from other carriers at much lower rates than those charged by the telephone company.

With the advent of competition in the telephone marketplace, it has become imperative for customers to be able to evaluate the various service and equipment offerings and to match them with their own business requirements on a cost-benefit basis. Choosing from the many services and vendors available is a difficult task. To make the right decisions, management must have comprehensive, up-to-date knowledge of telephone equipment, services, vendors, and prices. Effectively managed, the telephone system can make an important contribution to the profitability of any organization.

Organization of the Telecommunications Function

The relative position of telecommunications in the organization has changed considerably since the advent of competition. In days when the manager's responsibilities consisted primarily of ordering service and paying the bills, telecommunications was generally assigned as a minor responsibility to one of the main company departments. Now, however, management must be able to select the best equipment, services, and vendor(s) as well as exercise ongoing system controls, all in a highly competitive environment. As a result, the telecommunications function is becoming more important and consequently more visible. Many companies are addressing the increasing importance of telecommunications by establishing separate departments to administer this function. These departments are usually staffed with professionals trained in telecommunications.

The number of persons assigned to telecommunications is determined by the amount of work to be performed. This, in turn, is governed by the size of the organization. In a small organization, the telecommunications responsibilities may be performed by a manager or supervisor who is also responsible for many other tasks. If extensive changes in the telephone system are required, or if a new system is to be implemented, the small company would probably hire a consultant.

As a general rule, the telecommunications department budget should be 3 to 6 percent of the organization's telecommunications expenditures. This expense should be more than offset by the reduction in telecommunications costs. In addition, a professional manager should be able to effect improvement in the quality of telecommunications service.

An organization with an annual telecommunications expense in the $700,000 range needs a full-time communications manager, while an annual telecommunications expense of $2–$3 million would generally warrant three specialists—a manager, an auditor, and a systems analyst.

The Management Process

Management has been defined in many ways; there is no universally accepted definition. For the purpose of our discussion, the following definition of management will be used: *Management* is the process of directing the efforts of organization members toward stated organizational goals. The basic activities a manager performs are generally referred to as the "functions of management." They are:

1. planning—setting organizational goals and developing methods of achieving them
2. organizing—grouping activities, assigning activities, and providing the authority to carry out the activities
3. leading—directing and motivating people to perform tasks essential to goal achievement
4. controlling—setting standards, measuring performance against standards, and taking corrective action as required

Functions of Telecommunications Management

The goal of telecommunications management is to provide good telecommunications services for an organization and its employees at the lowest possible cost.

Telecommunications management functions include:

1. administering the ongoing operation of the telecommunications system
2. preparing and administering the telecommunications budget
3. keeping abreast of changes in equipment, services, industry structure, and rates
4. implementing and administering strategies for usage control and instructing company employees in efficient usage procedures
5. assisting top management in developing corporate telecommunications policy
6. planning and implementing changes required in the telecommunication system, including an entirely new system if warranted

The Telecommunications Manager's Responsibilities

Within the four general functions of management, the telecommunications manager is specifically responsible for:

- ☐ Planning
  - Setting departmental goals
  - Preparing departmental budgets
  - Reviewing present telecommunications system
  - Validating telephone bill(s)
  - Keeping current on technical, regulatory, and corporate changes
- ☐ Organizing
  - Interpreting company policy
  - Identifying projects
  - Writing job descriptions
  - Selecting and training personnel
  - Developing specifications for services
  - Choosing a consultant
  - Writing requests for proposals
  - Defining service standards
  - Assigning projects and tasks
  - Conferring authority necessary for performance
  - Allocating costs to users
  - Maintaining the company telephone directory
- ☐ Leading
  - Motivating people and developing team spirit
  - Guiding and developing personnel
  - Communicating with management and staff
- ☐ Controlling
  - Establishing measurement standards
  - Measuring service performance
  - Measuring cost performance
  - Measuring target date performance
  - Identifying and correcting deviations

Skills Required of a Telecommunications Manager

To effectively perform the responsibilities of a telecommunications manager, a number of skills are required, including:

1. a knowledge of vendors and services
2. an ability to translate an organization's telecommunications requirements into the most economical configuration of equipment
3. a knowledge of traffic engineering principles
4. a knowledge of the legal and regulatory aspects of telecommunications
5. the ability to understand tariffs and rate changes

6. proficiency in applying human relations principles
7. an ability to apply sound management techniques that cut expenses without radically increasing equipment
8. the skill to plan moves and changes and predict future equipment needs
9. a knowledge of usage control techniques
10. the ability to communicate effectively
11. a knowledge of budgeting principles
12. a knowledge of cost accounting procedures to allocate telecommunications costs to appropriate departments

The remainder of this chapter discusses the administration of an established telephone system. The following chapter will examine the selection and implementation of a new telephone system.

Managing the Established Telephone System

Telephone management is a service function. Effective management depends upon a thorough knowledge of all aspects of the system and a way to measure system performance.

Reviewing the Present System

The first responsibility of the manager is to become familiar with the existing telephone system. The manager personally visits each department of the company to verify the company's equipment records and discover how every telephone in the organization is used. During these visits the manager may locate equipment that is no longer necessary and can be removed from service. (Unfortunately, many departments request additional service as needed but neglect to request removal of unused equipment.)

These contacts establish communication between the manager and other departments and provide an opportunity to discuss their needs and solicit their suggestions. It also acquaints the manager with the company's business and the role that the telephone system plays in the successful operation of the business.

Identifying Systems Costs

Any study of telephone costs begins with a review of two types of documents: the telephone bill and the telephone equipment record. Both documents are available from the telephone company on a monthly basis.

The Telephone Bills One of the responsibilities of the manager is reconciling the telephone bills each month. To do this, the manager must understand how the bill is broken down. In the present post-divestiture era, users receive at least two bills: one from their local telephone company and another from their long distance carrier. The typical telephone bill (Figure 9.1) contains eight basic items, at most. Each item pertains to a separate portion of telephone cost. These items are:

1. local service
2. local messages
3. directory assistance, toll calls, and telegrams
4. other charges and credits
5. total of current charges excluding taxes
6. taxes
7. total of current charges including taxes
8. balance from previous month

Local service is generally the first item on the telephone bill. It represents the total charge for all local telephone service. Local telephone service is service to telephones in the same zone and certain adjacent zones. Local service is billed one month in advance.

The *local messages* item is the *total* charges for calls made within the local calling area (Figure 9.2). These calls are generally bulk billed; no call details are available. Flat rate service, which is available in some areas for residential use, is only available for business in small communities. (With flat rate service, this line on the bill is blank.)

Directory assistance, toll calls, and telegrams shows the *total* charge for these items. A separate statement is also supplied that provides call details. The statement shows the following information on toll calls: the date on which the call was made, the type of call (such as customer dialed, calling card, person-to-person, collect, conference,

ANY TELEPHONE COMPANY—BOX 1234, ANYTOWN, MICHIGAN, 48207
SER CHG FOR AUG 07–SEP 06 * DUE AUG 27, 1984, (313) 358–9876

ABC CORPORATION
12345 NORTH TWELFTH STREET
ANYTOWN, MI 48705

BUSINESS SERVICE MONTHLY CHARGE (IN ADVANCE)	132.82
LOCAL AND ZONE CHARGES (STATEMENT ENCLOSED)	651.73
DIRECTORY ASSISTANCE USAGE (SEE DETAILS)	.22
ITEMIZED CALLS (SEE STATEMENT)	290.58
U.S. TAX 33.42 STATE TAX 36.01 TOTAL TAX	69.43
TOTAL AMOUNT DUE	1,183.48

REFER BILLING OR SERVICE QUESTIONS TO (313) 968–1234

Figure 9.1
Sample Monthly Telephone Bill
The Business Service Monthly Charge does not include any charge for equipment because equipment must be either owned by the customer or provided by another vendor.

Figure 9.2
Local and Zone Call Detail Billing

LOCAL AND ZONE CALL DETAIL FOR (313) 358–9876

	CALLS	MINUTES	DISCOUNT	
LOCAL USAGE: 2,745 CALLS @ 8.2 CENTS				225.09
NEAR ZONE				
DAY RATE	1,656	5,946	00	386.27
EVENING RATE	149	499	9.80	22.87
NITE/WEEKEND	25	239	7.25	7.24
FAR ZONE				
DAY RATE	15	66	00	8.90
EVENING RATE	1	1	.05	.11
NITE/WEEKEND	2	19	1.25	1.25
TOTAL LOCAL AND ZONE CHARGES				651.73

Figure 9.3
Details of Itemized Toll Calls (Toll Billing provided by the local telephone company only itemizes calls that the local telephone company is authorized to handle. Other long distance telephone companies provide similar itemized statements for calls that are handled on their network.)

DETAILS OF ITEMIZED TOLL CALLS FOR (313) 358–9876

NO.	DATE	KEY	PLACE		AREA NUMBER	TIME	MIN	
1	7/11	D4	PONTIAC	MI	313 338 XXXX	1245PM	1	.12
2	7/13	D5	UTICA	MI	313 732 XXXX	324PM	9	.68
3	7/13	D4	MONROE	MI	313 289 XXXX	237PM	3	.93
4	7/14	D4	PONTIAC	MI	313 981 XXXX	359PM	6	.47
5	7/15	D6	MONROE	MI	313 459 XXXX	910AM	4	1.21
6	7/15	D4	PLYMOUTH	MI	313 349 XXXX	1135AM	1	.06

KEY: B-BILL TO THIRD NUMBER, C-COLLECT, D-CUSTOMER DIALED STATION, E-ENTERPRISE, F-CONFERENCE, H-CALLING CARD, J-CIRCLE CALLING (30% DISCOUNT OFF RATE IN EFFECT IN STATE), K-PERSON CALL BACK, L-CUSTOMER DIALED PERSON, M-MOBILE, P-PERSON, T-TELEGRAM, X-BUDGET TOLL DIALING (30% DISCOUNT OFF RATE IN EFFECT IN STATE), 4-INSIDE STATE (FULL RATE), 5-INSIDE STATE (EVENING 30% DISCOUNT OFF FULL RATE), 6-INSIDE STATE (NIGHT AND WEEKEND 50% DISCOUNT OFF FULL RATE).

mobile, ship to shore) the place called, the telephone number called, the time of day the call was made, the length of the call in minutes, and the total cost of each individual call (Figure 9.3). The statement also provides details of telegram calls that are handled by Western Union and billed through the telephone company. In addition, some long distance carriers charge for long distance directory assistance and include the total of these charges on the toll statement. The customer receives a separate toll statement from each long distance carrier that has been used.

The statement also provides details of directory assistance charges and telegrams (Figure 9.4). Directory assistance billing depends upon the charge plan of the serving telephone company.

Other charges and credits shows the *total* of all other charges and credits. A separate statement of these one-time charges and credits is supplied with call details. This section includes all charges that do not appear elsewhere, such as installations, moves, changes, and other adjustments.

Most telephone companies charge for equipment items and flat rate service in advance. This means that the first telephone bill will

Figure 9.4
Directory Assistance Billing

DIRECTORY ASSISTANCE USAGE DETAIL FOR (313) 358-9876	
211 CALLS TO 1-555-1212	
LESS 200 CALL ALLOWANCE	
11 BILLABLE CALLS @ 22 CENTS PER CALL	2.42
LESS MONTHLY CREDIT @ 22 CENTS PER LINE	2.20
NET DIRECTORY ASSISTANCE USAGE CHARGE	.22

include one entire month in advance plus the portion of the month in which the service was begun. Similarly, it will list a credit for any service discontinued during the month.

The Equipment Inventory The inventory as maintained by the local telephone company may be obtained upon request. The inventory contains a detailed itemization of the customer's equipment and circuits, including service features. The record usually is in coded form; codes will vary from telephone company to telephone company. Generally the record shows each item of equipment classified by telephone number and USOC (Universal Service Order Code) number. USOC codes were developed by the Bell System to provide uniformity in describing its services. Interpretation of the codes may be obtained from the telephone company representative.

Establishing Corporate Telecommunications Policy

A major responsibility of the telecommunications manager is controlling telephone expense. To do this, the manager establishes guidelines for the provision of telephones and controls on telephone usage. Without controls, telephone costs are almost certain to be unnecessarily high and continue to escalate.

The telecommunications manager is the key person in developing procedures and controls for telephone provision and usage. In developing these procedures and controls, the manager works closely with senior management. After approval by senior management, the procedures will become the cornerstone of corporate communications policy.

Procedures will be required for:

1. providing telephone equipment and services
2. controlling usage
3. allocating costs to system users
4. training system users

Providing Equipment and Services A basic principle of the telecommunications policy should be that all orders for telephone equipment or service be placed through the telecommunications manager.

Controlling Telephone Usage Telephone usage can be controlled either by manual or automatic (electronic) methods. One method is to require all calls for which a charge is made to be placed through an operator. The operator logs all such calls for later comparison with the telephone bill. Some companies modify this procedure by asking users to log their own calls; however, this provides a less reliable record.

As a general rule, placing calls through an operator acts as a deterrent to indiscriminate usage. However, if a more stringent procedure is desired, the operator can be instructed to ask whether the call was business or personal but to put it through irrespective of the answer. Just asking the question will generally cause users to think twice before placing personal calls.

Most modern telephone systems can be equipped with automatic station identification capability. With this feature, call details are automatically recorded: identification of calling and called stations, the time the call was placed, the length of the call in minutes, and the cost of the call. Thus, the cost can be allocated to the extension making the call. The feature also acts as a deterrent to unauthorized usage. Two such features, Automatic Identification of Outward Dialing (AIOD) and Station Message Detail Recording (SMDR), were described in Chapter 8.

Some modern telephone systems are available with a service feature that requires user identification codes. This feature requires the caller to enter an identification code before any long distance call can be completed. Caller identification codes are used to restrict usage by unauthorized callers.

Allocating Costs There are several important principles of cost allocation:

1. The communications manager is responsible for providing the most economical service that will meet the needs of the various departments of the company.
2. Each department must bear the cost of the service that it uses.
3. The communications manager is responsible for identifying and calling attention to opportunities for improving telecommunications cost performance.

The telecommunications manager should receive and process all invoices from telephone companies and telephone vendors. Each bill must be compared with the corporate inventory record. Any discrepancies discovered must be reconciled with the carrier and refunds or credits negotiated. This function must be performed for each location for which the manager is responsible. After the accuracy of the bill has been determined or discrepancies reconciled, the manager has two important responsibilities. One is to make sure that the bill is paid, and the other is to allocate as many of the charges as possible back to the department responsible for the expense.

Charges for equipment and services can be prorated among departments on the basis of the equipment in each department. For example, if the company is composed of five departments each of which has eight speakerphones, the expense for the speakerphones would be divided equally among the five departments. However, if one department has only four speakerphones while the others have eight, the portion of its expense for speakerphones would be only half as much as that for the other departments.

Charges for local messages and zone call units cannot be identified and therefore must be prorated on an arbitrary basis. Prorates can be computed on the basis of the percentage of company employees in each department or percentage of the company's telephones in each department.

Toll calls can generally be identified from operator or user logs or automatic call detail recordings. Where this is the case, they can be billed directly to the department where the call originated. Other billed items that cannot be directly attributed to specific departments may be prorated on the basis of department size.

Training System Users One of the manager's responsibilities is training system users. It may be accomplished by any or all of the following methods:

1. intracompany training programs
2. instruction pages in the company telephone directory
3. in-house training by the telephone company
4. intracompany letters or memos
5. bulletin board announcements

The telecommunications manager should develop an intracompany training program for new employees and assign responsibility for its execution. This program is often coordinated with the orientation program conducted by the personnel department.

The company telephone directory offers another means of instructing employees about the most efficient way of placing various types of calls.

The introduction of a new telephone system may require in-house training. Such training frequently consists of classroom sessions conducted by the telephone company or other vendor. Generally, written instructional material is available at these sessions.

Intracompany letters, handbooks, or memos are often used to define company policy for telephone usage. For example, company policy might require personnel away from the business location to direct-dial toll calls rather than use company credit cards. Reimbursement for such business calls might be handled through the employee's expense account. Bulletin board announcements can be used to announce new services or changes in usage practices.

Training activities must be conducted on a continuing basis in order to make effective use of the telephone system.

Controlling the Telephone System

Control is the process that an organization employs to ensure that its activities are going according to plan. A basic premise of management is that effective management requires control.

The department's goals and objectives are established in the planning process. The control process measures progress toward these goals and alerts the manager to deviations from the plan that might require corrective action.

The control process consists of four basic steps:

1. setting standards
2. measuring performance
3. evaluating performance against standards
4. taking corrective action

Control addresses the basic questions of: Where do we want to go? Where are we now? What actions are required to reach our goals?

The Role of Standards A standard is a benchmark or point of reference against which performance can be compared. Without standards, there is no basis for measurement. Standards define desired performance and serve as the manager's operational goal. They are established in the planning process and reviewed periodically so that they can be modified as conditions change.

Standards should be as precise as possible. Ideally, they should be defined in quantitative terms. Quantitative standards have two principal advantages. First, they are specific, defining performance in clear-cut terms. For example, a quantitative standard for the length of a business long distance call is often established at 5.2 minutes. This tells the manager what the target length of business long distance calls should be.

A second advantage of quantitative standards is their reliability or consistency of measurement. Quantitative standards can be measured objectively. Thus, if two or more persons were comparing actual practice with standards, the decisions or ratings of each would be the same. Similarly, if the same person compared the same condition at different times, the results would always be the same.

All standards should be stated as precisely as possible, but some standards are difficult to describe in quantitative terms. Because setting quantitative standards for all activities is quite difficult, qualitative standards are also used. Qualitative standards are subjective. They are therefore of questionable reliability, since reliability is directly affected by the degree to which judgments, biases, and emotions of the evaluator enter into the evaluative process.

Most telecommunication departments wish to provide "good service." However, "good service" is qualitative and subject to many interpretations. The current approach of managers is to try to translate qualitative standards into objective measures. For example, the goal specifying "good service" might be qualified to state, "good service as evidenced by no more than three blocked calls per hundred call attempts during the busy hour." *Busy hour* is the two consecutive half-hour periods of the day in which the largest number of calls occur.

The manager must decide which areas of telephone service to control. The areas selected should be those most critical to the attainment of the department's overall goals and objectives. Since the overall goal of telecommunications is to provide good service at the most economical cost, the telecommunications manager would probably concentrate on controlling:

1. system capacity
2. costs
3. usage
4. special services

In setting standards in these four areas the manager is guided by company policy. Most companies wish to provide sufficient service for efficient business operation. This policy is reflected in how they provide telephone service.

System Capacity Telephone systems are available in many different sizes and configurations, ranging in size from only a few telephones to very large systems with several thousand telephones. A major concern of the telecommunications manager is how effectively the existing system serves the needs of the organization. This, in turn, is determined by the number of users, the number of calls to be handled, and the grade of service to be provided.

The utilization rate of the system can be computed by comparing the number of telephones installed to the maximum number of telephones that can be connected to the system. Thus, if a telephone system with a physical capacity of 400 telephones had 368 telephones installed, the utilization rate would be 92 percent. The manager uses utilization data to monitor the adequacy of the system. If the utilization rate is consistently low, it indicates probable waste of capacity. Conversely, 100 percent utilization indicates that there is no room for growth. In this case, the manager must decide whether to supplement the system (if this is possible), obtain a new system, or fail to provide new service to anyone.

Just as telephone systems differ in the number of telephones that they can physically serve, they also differ in how many calls they can handle simultaneously. If more calls are generated at the same time than the system can handle, some calls will be blocked and the service will probably be unsatisfactory.

When the telecommunications manager selects a telephone system, a decision must be made about the desired *grade of service.* Grade of service is the probability of a call being blocked or delayed. It is expressed as a percentage. The better the grade of service, the more the equipment will cost. Thus, top management should be involved in selecting the grade of service that is being purchased. Generally, telephone companies engineer the public telephone network to provide P.01 grade of service. Private telephone systems can be engineered to provide nearly any desired grade of service. However, most private systems do not meet the P.01 standard.

Costs A budget is a plan describing how the financial resources of the organization will be used. It shows what portion of the total resources will be devoted to each expense item. The budget functions as a standard of financial control. Budget preparation and administration make up one of the manager's important responsibilities.

Preparation of the telecommunications budget starts with analyzing the telephone bill. The information contained in the telephone bill should be summarized in whatever way will best demonstrate the history of the departmental expense. One way to analyze telephone expense is to maintain monthly records of the bills, summarizing them as follows:

1. local services
2. local message charges
3. long distance charges (including WATS)
4. other charges
5. other carrier charges
6. charges for directory advertising
7. charges for installations, moves, and changes
8. credits
9. taxes

Figure 9.5 illustrates a form for summarizing monthly telephone expenses.

Figure 9.5
Summary of Monthly Telephone Expense

Date	Service	Message Units	Long Distance (Incl WATS)	Other Charges	Other Carriers	DIR Chgs	Moves & Chgs	Credits	Taxes	TOTAL

When these figures are posted monthly over a period of time, the resulting cost history will serve as a basis for budget preparation. In addition to the costs for telephone services, the manager must add the expense of the telecommunications department, including the cost for any personnel needed to support the system.

The manager must be able to document past costs to convince corporate management that the expenditures that the budget specifies are necessary and that the budgeted expense will produce the quality of service that management wants.

In addition to serving as a basis for budget preparation, the telephone cost history may reveal trends that require corrective action. For example, if the cost history shows that long distance costs are escalating rapidly, the manager might want to consider implementing more stringent usage controls or using other common carriers, WATS lines, FX lines, or private lines.

The control process involves comparing actual telecommunications costs with projected costs set forth in the budget. Any discrepancies between actual and budgeted costs should be noted and carefully analyzed. An expense that exceeds the budgeted figure is an *overrun*. Any manager whose department overruns the budget is faced with a difficult situation. There are only two solutions—to cut expenses or to add money to the budget. Neither of these solutions is entirely satisfactory. Therefore, the importance of a realistic budget cannot be overemphasized.

An underrun of the budget may appear to be easier to handle, but underruns can present problems too. When the comptroller or top management sees an underrun, the usual reaction is to conclude that the department's budget figures were too high and to expect the next year's figures to be reduced accordingly.

Costs are always of major concern to an organization. The high costs of communications as a percentage of the total business operating costs makes control of this area extremely important.

Usage It is an accepted fact that telephone bills are escalating rapidly. While the manager cannot control telephone company rate increases, substantial savings can result simply from reducing system abuse. It is the manager's responsibility to establish rules and procedures for the use of business telephones. These rules should include:

1. a company policy concerning the use of business telephones for personal calls—both local and long distance.
2. a company policy concerning special calls—person-to-person, calling card, charge to third party, etc.
3. a company policy concerning the provision of special equipment and services, such as speakerphones or WATS lines.
4. a continuing program for training telephone system users.
5. an effective procedure for allocating all telephone services and usage costs to the user.
6. a follow-up procedure for dealing with system abuse.

Monitoring both local and long distance usage is a continuing responsibility of the manager. There are specific actions that can be taken to reduce or eliminate unreasonable use. Probably the most effective procedure for controlling telephone usage is allocating costs to the users and making certain that the users are aware of this practice. Knowing that call details are being logged or recorded will do much to eliminate unnecessary usage and to shorten call duration.

The user training program can be a vehicle for educating system users about the various pricing mechanisms for telephone costs; for example, WATS line billing. When WATS service was initiated, the service was billed on a flat rate basis. This led many users to believe that if a WATS line was available, its usage would be "free." Although new WATS tariffs are based on usage, there are still people who regard WATS lines as a way to obtain "free" calls. Old ideas die hard (especially when people want to believe something), but education can do much to eliminate abuse based on misinformation regarding telephone costs.

Special Equipment and Services Technology has provided an abundance of new telephone equipment and services. While some of these features fill a real need, others may add prestige to an office but seldom be used. As the telecommunications marketplace continues to become increasingly competitive, we can expect to see more and more advertisements about these sophisticated new products and services. Once users are aware of their advantages, it is human nature to want these goodies. It is the responsibility of the telecommunications manager to evaluate service and equipment items realistically and to acquire them only when they will contribute to the total efficiency of the organization.

As we discussed earlier, the manager must verify the equipment inventory and evaluate the usefulness of special equipment. As changes are made in personnel, the manager should discuss special equipment with newcomers to be sure that the equipment will actually be used and will contribute to the effectiveness of the organization.

Some of the special equipment and service features available include speakerphones, conferencing, call forwarding, automatic (speed) dialing, automatic call detail recording, camp-on, identified ringing, message waiting, and paging. These features were described in detail in Chapter 8.

The Fully Utilized System

If feedback obtained during the control process indicates that the system is operating at full capacity or is already overloaded, the manager should study the alternatives available to remedy the situation.

The first consideration should be whether to enlarge the present system by providing additional equipment and trunks. Where this is possible, it is often the most practical thing to do. However, it might not be possible because the system may already be configured at capacity. Or the system may be obsolete so it is impossible to obtain additional equipment or parts compatible with the system.

Still another reason why it might not be possible to enlarge the present system might be the lack of physical space. If the room where the equipment is housed is filled, rearrangement to include more equipment is impossible.

When it is possible to enlarge the present system, the manager must investigate the feasibility of doing so. Even when it is possible to add to a given system, it may or may not be the best thing to do. The manager should evaluate conditions regarding the system before making a decision. One such condition is the commitment to the vendor. Where equipment is leased, as is often the case, the length of time left to run on the existing contract is an important factor. When the contract with the vendor is soon to expire, any necessary changes should be considered before an organization signs a new contract. Once a contract has been signed, there is usually a substantial penalty for terminating it.

If the system is owned by the organization, the possibility of its reuse at another company location or its possible sale can be considered.

Future company plans must also be taken into account. If the company plans might affect the decision, they must be considered. For example, if company plans include moving the offices to another location, this certainly has a bearing on any decision regarding the enlargement of the telephone system.

Because telecommunications technology, regulation, and supplier offerings are changing at an incredibly fast pace, communications decisions are becoming increasingly complex. The effective manager must have regular briefings on these important aspects of the business in order to provide sound telecommunications management.

This discussion has described telecommunications management in the context of providing a service to an organization and its employees. Efficient management of the telecommunications system entails providing the best possible service at the least possible cost.

When telecommunications service is being provided for external use—that is, for users who are not members of the organization—telecommunications can be managed not only to provide efficient service but also to generate a profit. Examples of this approach include hotels, hospitals, and school/college dormitories.

Summary

Prior to the advent of competition, most organizations relied upon the telephone company to manage their telephone services. After the competitive market developed, it became necessary for customers to evaluate the service and equipment offerings of various vendors and match them with their own requirements. Choosing among the proliferation of services and vendors available is a difficult task, requiring a comprehensive knowledge of telecommunications and the application of sound management principles. Effectively managed, the telecommunications system can make an important contribution to the profitability of any organization.

The goal of telecommunications management is to provide good telecommunications services for an organization and its employees at the lowest possible cost. Effective management of the telecommunications system requires a thorough knowledge of all aspects of the system and a way to measure system performance.

The first responsibility of the manager is to become familiar with the existing system, including system inventory and service and equipment costs. The manager should be a key person in the establishment of corporate telecommunications policy. Policies are required for:

1. providing telecommunications equipment and services
2. controlling usage
3. allocating costs to users
4. training system users

Another responsibility of the manager is to formulate standards for system capacity, costs, usage, and special telephone equipment. The manager monitors system performance and measures it against the established standards. When data obtained through the control process shows that the system is operating at or near capacity, the manager explores alternatives to remedy the situation.

Review Questions

1. What is the objective of telecommunications management?
2. Historically, many organizations assigned the responsibility for telecommunications service to an office manager or clerical supervisor. Why was this possible?
3. What factors have been instrumental in bringing about the need for telecommunications management?
4. What are some of the options an organization has today in obtaining telecommunications equipment and services?
5. What are the functions of telecommunications management?
6. What is the first responsibility of the telecommunications manager?
7. What are the basic areas for which telecommunication policies are required?
8. What are the skills, knowledge, and understanding required of a telecommunications manager?
9. How does a telecommunications manager evaluate system adequacy?
10. What is telephone cost history data used for?
11. What are some of the procedures a manager can implement to control telephone usage?
12. Why might it not be possible to enlarge an existing telephone system?

References and Bibliography

Arredondo, Larry A. *Telecommunications Management for Business and Government,* 2d ed. New York: Telecom Library, Inc., 1983.

Datamation. "Dialing Dilemmas: How Telecom Managers Are Learning to Cope with the AT&T Divestiture." *Datamation,* January 1984, 118–24.

Inman, Virginia. "Busy Signal: Corporation Managers of Telecommunication Find Job Gets Tougher." *The Wall Street Journal,* December 28, 1983, 1, 6.

Kaufman, Bob. *Cost-Effective Telecommunications Management.* Boston: CBI Publishing Company, Inc., 1983.

Kuehn, Richard A. *Cost-Effective Telecommunications.* New York: AMACOM, A Division of American Management Association, 1975.

______. *Interconnect: Why and How,* 2d ed. New York: Telecom Library, Inc., 1982.

Nabs, Charles N. "Strategic Planning for Telecom Management in the 1980s." *Business Communications Review,* Vol. 13, No. 3 May–June 1983, 12–18.

Newton, Harry. *Professional Management via Telecommunications,* reprints from *Business Communications Review.* New York: The Telecom Library, Inc., 1980.

Price, Margaret. "Seizing the Moment: Savvy Execs Find New Leverage." *Industry Week,* April 2, 1984, 48–51.

Reynolds, George W. *Introduction to Business Telecommunications.* Columbus, Ohio: Charles E. Merrill Publishing Co., 1984.

Rue, Leslie W., and Lloyd L. Byars. *Management Theory and Application,* revised ed. Homewood, Ill.: Richard D. Irwin, Inc., 1980.

Self, Robert L. *Long Distance for Less,* 2d ed. New York: The Telecom Library, 1983.

Selwyn, Lee. "Managers Face Challenges in New Industry Created by Deregulation and Divestiture." *Communications News,* July 1983, 66D–66H.

Stoner, James A. F. *Management,* 2d. ed. Englewood Cliffs, N. J.: Prentice-Hall, Inc., 1982.

Vonarx, Mark. "Communications Options Cut Costs." *Word Processing and Information Systems,* August 1982, 24–29.

Selecting and Implementing a New Telephone System

10

A few years ago people would have laughed at the thought of "shopping" for a telephone system. A business needing a new system called the local telephone company representatives and left most of the decisions up to them. The manager's role was limited to casual concerns about details of the new system. Today, however, users can shop for telephone equipment from both the local telephone companies and interconnect companies. Businesses can select their own telephone system, "shopping around" in much the same way as they would for any other major purchase.

Many companies are convinced that they can save money by owning their own phone equipment rather than paying a monthly rental fee to the telephone company. Of course, organizations still pay the phone company every month for their phone calls, but they are relieved of monthly phone equipment bills. In any case, the management of an organization must now make informed decisions regarding telephone equipment, service, and financial arrangements. Thus, the job of the telecommunications manager is more complex than it formerly was.

Telecommunications management is charged with providing telephone service that meets the needs of the organization. This requires continuing analysis and evaluation of the system's capability to meet service demands. Details of the analysis will identify possible shortcomings of the present system and help determine whether to improve, supplement, or replace the system.

There are three situations in which a new telephone system may be required:

1. to provide service for a new location
2. to replace a system that lacks capacity to meet the needs of the organization

3. to replace a system that cannot provide desired service features, such as direct inward dialing or touchtone capability for communication with computers

This chapter describes the options available to an organization selecting a new telephone system and will guide the reader through the series of tasks necessary for successful implementation of the system.

Selection of a Telephone System

The selection of a new telephone system is one of the most important, and probably one of the most difficult, responsibilities of the telecommunications manager. It is also one in which the manager has generally had little or no practice. The proliferation of vendors, each claiming to have the "best" system; the many attractive service features available; and the variety of financial arrangements available present the telecommunications manager with a diverse set of choices and make the selection more difficult.

There are three options available to organizations for acquiring a new telephone system: purchase, lease agreement, or rental agreement. Regardless of which option the organization chooses, it makes a long-term commitment of large sums of money, thereby precluding the opportunity to take advantage of further system advances for the duration of the commitment.

Rapid advances in telecommunications technology have resulted in a wide variety of improved service features, often at reduced costs. These advancements tend to build obsolescence into existing telecommunications equipment. Most organizations acquiring new telephone systems want the new system to have the latest capabilities. The implicit threat of system obsolescence poses a dilemma for the manager and acts as a deterrent in the decision-making process, which further complicates the manager's job.

Role of Vendors in System Selection

The selection of a new telephone system must be based upon a careful determination of the organization's needs, extensive review of the various service offerings, and accurate projections of future needs. An examination of the system analysis procedures followed by the telephone companies prior to divestiture will be helpful in understanding the process of selecting a new telephone system.

Telephone company marketing representatives served as the liaison between the telecommunications manager and the telephone company. They were trained in the communications needs of a particular industry. Very large nationwide organizations often required the services of many marketing representatives, headed by a national account manager. The Bell System's marketing department divided its customers into three broad categories:

1. government, education, and medical
2. commercial
3. industrial

These categories were further broken down into various industries within the categories. For example, the commercial category consisted of such firms as banks, law firms, automobile dealers, and department stores. Similarly, the industrial category included all types of manufacturing organizations.

Marketing representatives were assisted by traffic and equipment engineers who translated usage data into quantities of equipment and trunks and specific technical design details.

Traffic Studies The telephone company conducted traffic studies to help an organization determine its telephone needs. These studies were used to determine the adequacy of the existing telephone system and provide a basis for the design of a new system.

Traffic studies consist of counts of busy-hour calls classified by types of calls, such as intracompany calls, outward local calls, inward direct-dialed calls, and operator-handled inward and outward calls. Each of these counts is divided by the number of telephones in the system to determine the *calling rate,* or number of calls per telephone, for that particular type of call. The calling rates of the existing telephones are multiplied by the estimated numbers of telephones to be served by the new system. The resultant numbers of calls are used by traffic engineers to determine types and quantities of equipment and trunks that will be required in the new system. For example, if the existing system receives 200 incoming calls during the busy hour and there are 100 telephones in service, the inward calling rate for the system will be 2.0 inward calls per busy hour per telephone. If the new system will have 400 telephones, 800 busy-hour inward calls can be expected, and the system must be engineered to handle this volume of incoming traffic. Similar calculations are made for each type of call handled by the system in order to determine total system needs.

Traffic studies also include counts of calls that cannot be completed because of a network busy condition. These calls are known as *overflows* or *blocked calls.* Grade of service is the ratio of blocked calls to attempted calls, expressed as a percentage. Thus, if the count of blocked calls is 20 and attempted calls 400, the grade of service would be 20/400 or P.05. This indicates a 5 percent overflow. The higher the percent of overflow, the poorer the grade of service. If the overflow is excessive, more trunks will be required. If more trunks cannot be added to the existing system, a new system must be obtained.

Another facet of traffic studies concerns long distance services. Traffic studies analyze long distance records to determine whether an organization does a substantial volume of calling to a locality that is not within the local calling area. Where this is the case, there may be a need for special long distance facilities, such as WATS, foreign exchange, or tie lines, and the system should be designed to include this capability.

As a result of the traffic study, the marketing representative was able to match system requirements with available types of systems and make an appropriate recommendation to the telecommunications manager. Each system was customized to meet the specific needs of the user organization.

In the post-divestiture marketplace, all vendors of telephone systems and services provide assistance in the form of traffic studies and specific recommendations similar to those formerly provided by the telephone company.

Role of the Telecommunications Manager in System Selection

The major distributors of telephone systems (other telephone companies and interconnect vendors) have patterned their organization after the Bell System concept of industry specialization. As a result, the telecommunications manager deals with many vendors in addition to the telephone company, each competing for the organization's business. Their objective is to convince the telecommunications manager that their telephone system best meets the organization's needs.

In the new environment of unregulated competition, the manager receives just as much assistance as formerly. Now, however, much of the advice received is conflicting, and nearly all of it is prejudiced in favor of the providing vendor. The manager should learn as much as possible from the various vendors in order to evaluate the sometimes conflicting recommendations and make the best decision.

Today the evaluation of an existing telephone system begins with a traffic study made by the telephone company, by interconnect vendors, by consultants, or by the telecommunications manager (Figures 10.1–10.3). A good policy is to obtain traffic studies from several sources so that the various recommendations can be compared and evaluated. The telecommunications manager often begins the

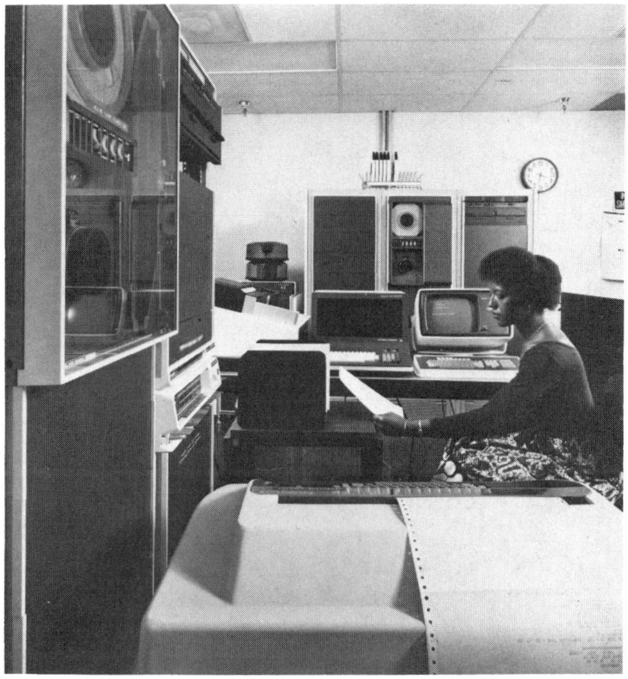

(Courtesy of ROLM™ Corporation)

Figure 10.1
ROLM Analysis Center™
The Analysis Center offers a wide variety of report formats for subscribers' call details.

(Courtesy of ROLM™ Corporation)

Figure 10.2
ROLM Telecommunications Analysis Center Library (TACL)™

Figure 10.3
ROLM CBX Analysis Center™
CBX Analysis Center™ reports provide managment with timely information to spotlight abuse, allow accurate cost allocation, and facilitate client billing.

(Courtesy of ROLM™ Corporation)

selection process by requesting a traffic study from the telephone company and/or other vendors.

Sources of Management Information In addition to a basic knowledge of telecommunications concepts, the manager must be familiar with the latest technologies and vendor service and equipment offerings. Telecommunications technologies are changing at an incredibly fast rate. The manager's education must be continually updated to keep pace with new developments. Some of the sources of management information include periodicals, trade associations, seminars, and consultations with vendors.

Periodicals, seminars, and trade associations can provide the manager with information on state-of-the-art technology and industry and regulatory developments. Consultations with vendors provide the manager with information on each vendor's service and equipment offerings. A list of trade associations, vendors, and organizations that sponsor seminars is found in Appendixes A through C.

Role of the Consultant

Given the fact that communications decisions have become increasingly complex with advanced technology and myriad vendors, an outside consultant can often save an organization substantial time and money in system selection. The function of the consultant is to give unbiased advice. A good consultant should be well informed on state-of-the-art technology, knowledgeable about all types of available systems, and honest and impartial in system evaluation and recommendations.

The consultant's contract should detail the work to be done and the fee to be charged. Organizations should expect to pay a fair daily rate, which can be upwards of $400 a day. Managers should avoid retaining a consultant whose fee is contingent upon cost-saving; anyone can reduce telephone costs by downgrading services. Organizations have a right to expect a consultant to improve service and/or save money.

The consultant's report should provide:

1. an evaluation of the existing system
2. new system requirements and specifications
3. alternative solutions for a new system
4. conclusions and recommendations

As a general rule, the consultant should stay on the job until installation has been completed and the system checked to see that it has been installed according to specifications.

Determining System Requirements

The next task after the traffic study is to determine the specifications for the new system. The areas to consider include:

1. Line capacity. How many telephone lines will be required over the lifetime of the equipment?
2. Call capacity. How many calls will the system be required to handle during the lifetime of the system?
3. Service features. What service features best serve the organization's needs and contribute to overall profitability?
4. Costs. What are the cost considerations in selecting a new telephone system?
5. Financial arrangements. Should the system be purchased, leased, or rented? What are the terms of the various financial operations?
6. System maintenance. Are diagnostic procedures to be included in the system? What are the provisions for system repair?

There are two types of physical capacity for any telephone system: line capacity and call capacity.

Line Capacity The physical line capacity is the maximum number of telephone lines that can be served by the system. For example, if the physical capacity of the system is 1,000 lines, it would not be possible to connect 1,001 lines to the system.

The telecommunications manager maintains records of the actual number of lines in service on the present system and combines this data with knowledge concerning the projected growth of the organization. These two sources of information are used to prepare the estimate of the telephone lines required. When the existing line capacity has been exhausted, the organization requires a new system.

Call Capacity Call capacity is the ability of the system to handle a specific number of telephone calls in the busiest hour of the day and still provide a specified grade of service. The manager must study call volume data on a continuing basis. If the busy-hour call volume exceeds the designed call capacity, calls will be blocked because of the equipment overload, and quality of service will deteriorate. The manager also uses call volume data to predict when call capacity will be reached and to estimate future requirements.

Sometimes call capacity can be increased by patching up the system, adding more trunks or switching equipment. Generally, however, a new system is required. Call counts are used to predict trends of future usage. These projections, in turn, are used to estimate the call volume that the system must be engineered to handle over its lifespan.

Service Features As has been noted, there are many service features available with modern telephone systems. A partial list of these features includes least-cost routing, direct inward dialing, automatic caller identification, toll restriction capability, call transfer, call forwarding, speed dialing, camp-on, and group pickup. Some of these features will produce cost savings, and others are just "nice to have." Research has shown that many of the "nice-to-have" features are, in fact, seldom used.

All digital telephone systems have certain switching and transmission advantages over older analog systems. Electronically controlled, programmable systems often can be upgraded to provide new service features. The need for some of these features could justify a new telephone system. All system features are not available from all vendors. Therefore, the choice of vendor may depend upon the availability of desired service features.

Costs Telephone costs represent a substantial and ever-increasing portion of an organization's budget. There are two underlying reasons for rising telephone costs. One reason is the higher telephone rates. Telephone service and equipment, like most other elements in our economy, have been affected by inflation. Another reason is increased telephone usage. Today, many multinational corporations have established branch offices in widely dispersed locations, with the resultant increase in communications requirements. In addition, businesses keep finding more uses for the telephone. Telemarketing, to cite just one example, has greatly reduced the volume of personal sales visits, but has caused a corresponding increase in telephone usage. While total telephone costs generally rise as an organization's size and activities increase, the objective in obtaining a new telephone system is to reduce the costs per employee or per unit of production.

Opportunities for cost savings in a new telephone system exist primarily in the area of service features. There is no opportunity for cost savings in the area of tariffed items, such as telephone trunks, provided by an operating telephone company. There is also very little, if any, opportunity for savings on telephone lines. Essentially, if an organization requires a specific number of telephone lines, that is the number of telephone lines that must be obtained.

The majority of new telephone systems are electronic rather than mechanical. Electronic systems combine computer capabilities and switching power to improve the performance of the telephone system, thereby enabling them to provide many service features unavailable in older telephone systems. Electronic systems can:

- route both incoming and outgoing calls over the least expensive route available
- dial numbers faster and more accurately than humans
- record details of telephone calls
- restrict unauthorized toll call usage
- identify system malfunctions, thus facilitating system repair

These capabilities can provide substantial cost savings; they are combined in various ways in vendor system offerings.

Financial Arrangements For many years, the only financial option available to a system user was rental from the telephone company. The telephone company tariff provided for a one-time installation charge and a monthly charge for equipment rental and service. Today, telephone systems may either be rented, leased, or purchased.

Under a rental agreement, the user receives telephone service for a monthly fee. The provider of the service is responsible for such things as maintenance, taxes, and insurance. In addition to the monthly service fee, the user pays a one-time installation charge and also pays for any moves, changes, or rearrangements in the system. Rental agreements are made with any vendor of telephone equipment. The user signs a contract that specifies the terms and conditions and provides for a substantial penalty for cancellation of the service.

Many interconnect vendors sell their systems outright. When a system is sold, the vendor receives full payment. The purchase price includes the cost of installation. The vendor has no responsibility for maintenance beyond the system's warranty. Many vendors offer service contracts that provide system maintenance for a specified fee. These contracts generally include an escalation clause to protect the vendor against rising costs. The owner of a telephone system is, of course, responsible for insurance and taxes and is eligible for depreciation and investment tax credits.

Another financial arrangement is leasing. To lease a telephone system, the organization enters into a contract with the vendor in which the organization agrees to pay a stipulated fee periodically in return for use of the system. The lease agreement specifies all the details of the arrangement, including the period of the lease, cancellation penalties, options to purchase, maintenance provisions, tax and insurance responsibilities, and depreciation and tax investment credits.

Many vendors only sell telephone systems; they do not offer lease arrangements. When this is the case, organizations often obtain a lease through a third-party agreement. Financial institutions and leasing companies are in the business of providing financial services; they make a profit by purchasing equipment and leasing it to users.

The telecommunications manager should prepare an analysis of the financial impact of each of the three options on the organization's financial status. This financial analysis must be examined within the framework of corporate financial policy in order to select the option that is best for that particular organization.

System Maintenance An organization that rents its telephone system obtains system maintenance as part of the rental agreement. An organization that purchases or leases a telephone system is responsible for providing maintenance either through a service contract or by its own maintenance staff. When an organization elects to perform its own maintenance service, personnel must be trained in system maintenance. Such training is generally available from equipment vendors. The training prepares the trainees to accomplish system moves, changes, and rearrangements as well as maintenance.

An advantage of modern electronic systems is their capability to identify defective system components either from remote locations or on site. When potential system malfunctions are identified, the user can replace defective modules or circuit boards before system breakdown occurs.

Competitive Bidding

Request for Quotation When the decision to obtain a new telephone system has been made, the telecommunications manager prepares a Request for Quotation (RFQ). The RFQ describes the type of equipment desired; the approximate quantity and types of telephones required; the approximate number of incoming and outgoing trunks; special features desired; maintenance arrangements; requested service date; and any other pertinent information.

The RFQ asks vendors to provide an estimate of the cost of the system, alternative financial terms, and a description of the standard service features of the proposed system. The RFQ should also specify the deadline date for submitting quotations.

Although the ideal procedure is to issue an RFQ, occasionally the need for the new system is urgent and it becomes necessary to bypass this step in the selection process and go directly to the next step, issuing the Request for Proposal.

Request for Proposal In securing a new system, it is generally desirable to obtain price quotations from several vendors as well as the telephone company. A convenient way to solicit such quotations is to issue a Request for Proposal (RFP), a formal document sent to potential vendors inviting them to submit bids on the new telephone system. The RFP must be very specific concerning the types and quantities of equipment required for the telephone system so that proposals from vendors will be realistic. The following items are usually included in system specifications:

1. System capacity: number of incoming, outgoing, and internal calls the system must be capable of handling during the busy hour.
2. Equipment: number of trunks (by types) required; number of telephones (by types) to be served by the system.
3. Service features: types and quantities required.
4. Changes: move and rearrangement capabilities.
5. Maintenance: diagnostic/repair capabilities of system and availability of maintenance service and parts.
6. Training personnel: training for system users; training on repairs, moves, and changes.
7. Proposal submission date: deadline for return of bids.
8. Financial options: rental, purchase, direct lease, or third-party lease.

Appendix D shows a sample RFP. This sample is for illustrative purposes and is general in nature; it should not be used in a specific situation. The RFP should be customized to meet the organization's requirements and approved by legal counsel before it is submitted to vendors.

Evaluating Vendors' Proposals The RFP responses should first be screened to be sure that they meet the system and service requirements of the organization in all respects. If they do not meet these requirements, they should be eliminated.

If all other things were equal, the proposals meeting system requirements could be ranked by price, with the one having the lowest price selected. However, no two organizations are identical in financial resources, credit ratings, cash flow, standards, and priorities. Thus, each proposal should be evaluated in terms of the organization's policies, preferences, and financial status. There are four basic factors to consider in evaluating vendors' proposals:

1. product
2. vendor
3. total system costs
4. financial options

Product When competition first entered the telephone marketplace, the service and equipment offerings of the interconnect vendors varied widely. Older telephone systems were chiefly mechanical. The development of solid-state electronics made a variety of new service features possible. Today all new telephone systems have essentially the same capabilities. As a result, nearly any service feature can be obtained from any vendor. The principal differences between telephone systems lie in the design arrangement of their components and in the physical characteristics of their telephone sets.

Different manufacturers arrange the functional components in different configurations, particularly in the number of components to serve the different types of trunks. The grouping arrangements must be matched with the requirements of the user. For instance, a new system requiring 12 trunks of a given type might have to be matched with manufacturers providing trunks in groups of 8, 10, or 12. Thus, the user requiring 12 trunks can select from one group of 12 trunks, two groups of 8 trunks, or two groups of 10 trunks. Since one group of 12 trunks would provide no expansion capability, this option can be ruled out. Therefore, two groups of either 8 or 10 trunks would be required. Since two groups of 8 trunks would provide sufficient expansion capability, this choice would be the most cost effective.

Telephone sets provided by different vendors also differ in physical characteristics such as color, weight, and shape. The aesthetic qualities of the telephone system should be evaluated in terms of the organization's preferences and priorities. Another factor that should be considered in evaluating the product is reliability. Product reliability is measured by the number of times the system is out of service during a stated time interval and the average length of such occurrences. This information is best obtained from other users of the system under consideration.

Vendor Factors used to evaluate the vendor include reputation, support staff, experience, location, and repair service. These factors are best evaluated by questioning existing customers of the vendor's system. Some questions to consider are:

- ☐ Does the vendor have a favorable reputation in the industry?
- ☐ How long has the vendor been in business?
- ☐ Is the vendor financially sound?
- ☐ Does the vendor have a support staff of highly competent marketing representatives, customer service engineers, and training personnel?
- ☐ Is the vendor's system marketed directly in the locality? In other words, can the organization deal directly with the vendor or must it deal with a third party to obtain the vendor's system in its locality?
- ☐ How does the vendor provide maintenance service?
- ☐ Are diagnostic and preventive maintenance procedures effective in avoiding serious system breakdowns?
- ☐ What is the maintenance service response time?
- ☐ What training does the vendor provide for system users?

Total System Costs Total system costs are the amount an organization has to spend to get the system it needs. These costs include such items as purchase price, total rental costs, installation, maintenance, interest charges, depreciation, and investment tax credits.

Financial Options Companies either purchase, lease, or rent telephone systems depending upon several factors, including the company's policies, preferences, and financial status. The telecommunications manager should make a detailed financial analysis comparing total system costs of the various financial options. Appendix E illustrates a telephone system financial analysis comparing rental, purchase, and lease options.

Under a rental agreement, the user pays a monthly equipment rental charge, which is subject to an excise tax. Additionally, the renter pays a one-time installation charge. The vendor pays taxes, insurance, maintenance, depreciation, and other expenses. The renter of a system can anticipate a yearly increase in rental costs because of inflation.

The advantages of renting are:

1. No large capital outlay is required.
2. Companies can break a rental agreement with relative ease—an important advantage in view of rapid technological developments.

The chief disadvantages of renting are:

1. Equipment choices are limited in rental systems.
2. Rates continue to escalate.

In buying a telephone system, the purchaser pays a one-time purchase price, which is generally subject to a sales tax. Because the purchaser of a telephone system is responsible for system maintenance, maintenance charges, adjusted for inflation, must be included in the total costs. In addition, the purchaser is responsible for insurance and taxes. The purchaser, however, is entitled to an investment tax credit for the first year and an annual depreciation tax credit for the depreciable life of the system.

The principal advantages of purchasing a telephone system include:

1. the avoidance of continuing monthly payments that can amount to several times the purchase price over a period of years
2. the protection from continuing rate increases
3. the realization of substantial investment and depreciation tax credits

The disadvantages include:

1. the large capital outlay
2. the possibility of owning outdated equipment
3. the difficulty of moving the equipment if the organization relocates

Under a leasing contract, the leasing company provides a telephone system for the organization in return for a specified monthly lease factor. This factor is determined by totaling all costs incurred by the leasing company—including insurance, taxes, and interest charges—and dividing this total by the number of months in the leasing contract. The system user is responsible for maintenance costs. In Appendix E , the user had the option to purchase the system after five years at 10 percent of the original purchase cost. After purchasing the system, the user is responsible for taxes and insurance.

Leasing a telephone system is similar in many ways to purchasing. The principal difference is that the large, immediate capital outlay is replaced by continuing monthly payments, with interest. An important advantage of leasing is that the lessee is protected from continuing rate increases. The principal disadvantage of leasing is that there is usually a substantial penalty for terminating the lease before its expiration. In most lease agreements, the lessee receives title to the system for a nominal sum of money after leasing it for a specified period of time, usually five years.

The financial study takes all expenditures in each alternative by periods (years), totals these amounts, and by use of discount tables calculates the present value of the monies spent in each year. This permits the manager to compare each of options and to determine the most attractive one from an economic standpoint. The analysis also shows the effect of taxes on costs for each option. Microcomputers using electronic spreadsheet software make it possible for managers to compare the various financial options.

The financial analysis of total system costs should be reviewed with the organization's financial officers so that this expenditure can be evaluated within the framework of the organization's total financial picture.

Implementing a New Telephone System

In the current deregulated environment of the telecommunications industry, the buyer and seller of a telephone system should enter into a written contract as a protection against possible misunderstandings. The contract should specify all details of the agreement, including the rights and obligations of each party. Prior to industry deregulation, the public utility commissions protected both customers and providers of services by approving tariffs that defined terms, conditions, and rates for all telephone services. The commissions were empowered to enforce tariffs and to settle disagreements. Under deregulation, however, the public utility commissions have no power to settle disputes; disagreements or performance defaults must be settled between the parties or in the courts.

Contract Provisions

The contract to acquire a new telephone system should contain provisions about:

1. System capacity. This specifies the capability of the system to handle a specified number of calls in the busy hour.
2. System configuration. This specifies the quantities of telephone sets by types.
3. Registration requirements. This specifies that the system must conform to FCC requirements regarding direct connection to telephone company lines.
4. Building code conformance. This specifies that all installation conforms to local building codes.
5. Warranty. This contains explicit details of the system's warranty.
6. System failures. This contains the seller's guarantee for restoration of service following system failures.
7. System testing. This specifies the types of and times for system testing prior to system acceptance and authorization of payments.

8. Operating manuals. This specifies the number of operating manuals to be supplied to the user organization.
9. Installation schedule. This specifies the schedule for completion of each event comprising installation, including cutover date.
10. User training. This specifies the vendor's responsibility to train system users in the system's service features.
11. Payment. This breaks the purchase price down into a series of payments that keep pace with system installation. It is customary to withhold a percentage of the total payment until system acceptance.
12. Other charges. This specifies the vendor's responsibility for such charges as freight and insurance until the title passes.
13. Title. This specifies when title passes to the buyer of the system.

This list is not intended to be all-inclusive but to suggest the major items that should be included in the contract to protect the buyer's interests.

System Cutover

The term *cutover* describes the process of putting a new system into operation. Before a new telephone system is ready to be implemented or cut over, a number of procedures must take place. They include:

1. Assigning telephone numbers. Each telephone number must be associated with a cable and wire pair assignment and the assignment details recorded. Wiring diagrams to show key telephone set assignments must be prepared.
2. Assigning user authorization codes. When the system incorporates user authorization codes, it is necessary to assign an individual identification code to each authorized user.
3. Selecting and implementing a computer program to allocate costs. The call detail data obtained from the telephone system must be interpreted so that each charge can be allocated to the person making the call. A separate computer program is required for this purpose.
4. Verifying the least-cost routing program. The system must be tested to ensure that the least-cost routing program is in operation.
5. Preparing traffic study procedures. The call count registers must be tested and appropriate data summarization forms prepared.
6. Preparing a system telephone directory. The names of all system users should be listed alphabetically, along with accompanying telephone numbers. Listings by departments should also be pro-

vided. The directory should also include detailed instructions for system usage.

7. Providing system training. System training should be provided for console attendants, system users, and personnel who may perform maintenance, moves, and changes.

The system cutover is scheduled for a specific time and date. The cutover should be a *turnkey* operation; that is, ready to go on an instantaneous basis at the turn of a key.

Experience has shown that call volumes on the first few days following a cutover are abnormally high because of curiosity calling. Therefore, judgments regarding system capacity should be withheld until a later date when the system has had a fair trial.

Summary

There are three options available to organizations for acquiring a new telephone system: purchase, rent, or lease. Regardless of the option chosen, the organization makes a long-term commitment of large sums of money that precludes the opportunity to make changes without incurring a substantial penalty. The implicit threat of system obsolescence makes the choice difficult.

In addition to a basic knowledge of telecommunications concepts, the manager must be familiar with the latest technologies and vendor service and equipment offerings. A basic task in selecting a new telephone system is to determine the specifications for the system. Traffic studies are used to help identify system requirements. The areas to be considered include line capacity, call capacity, service features, costs, financial arrangements, and maintenance provisions.

The first step in obtaining a new system is to prepare a Request for Quotation describing the desired system in general terms. This requests vendors to provide an estimate of costs and financial terms of the system.

The next step is to request a proposal. The Request for Proposal document is very specific concerning the types and quantities of equipment required for the system. An important part of the selection process is the financial analysis comparing rental, purchase, and lease options. The evaluation of vendors' proposals is a critical step in system selection.

The contract to acquire a new system should contain provisions about system capacity, system configuration, FCC registration requirements, building code conformance, warranty, system failures, system testing, operating manuals, installation schedule, user training, payment, and transfer of title arrangements.

The system cutover is scheduled for a specific time and date. It will be necessary for the manager to plan and follow an orderly sequence of procedures in order to effect an instantaneous cutover of the new system.

Review Questions

1. What are the factors that make the selection of a new telephone system difficult?
2. What was the role of the telephone company marketing representative in system selection?
3. What is a traffic study? How are traffic studies used in designing a new telephone system?
4. Why must the manager's telecommunications education be continually updated?
5. What are sources of management information?
6. What are the considerations in determining specifications for a new system?
7. How does a request for quotation differ from a request for proposal?
8. Before evaluating a vendor's proposal, what preliminary screening should be conducted?
9. What are the details that should be specified in the contract to acquire a new telephone system?
10. What are the procedures that must be completed prior to the cutover of a new telephone system?

References and Bibliography

Chorafas, Dimitris N. *Telephony: Today and Tomorrow.* Englewood Cliffs, N.J.: Prentice-Hall, 1984.

Darden, William E. III. "So You Want to Buy a New Telephone System?" *Business Communications Review,* January–February, 1984, 34–36.

Fiamingo, Josephine S. "All About Electronic Spreadsheets." *Office Administration and Automation,* February 1984, 44–48.

Holland, James W. "Choosing the Right Business Telephone System: Higher Efficiency, Lower Costs." *Telemarketing,* November–December 1982, 30–33.

Llana, Andres, Jr. "Basic Lessons on Telecom System Management for Users Buying Their First System." *Telephony,* July 18, 1983, 152–54.

Newton, Harry. "Long Distance: How to Survive." *Teleconnect,* August 1983, 26–29.

11 Principles of Traffic Engineering

In our mobile society the word *traffic* brings to mind automobiles and other channels of transportation. The term *traffic* is not limited to transportation, however. As used in communications, traffic denotes the flow of messages through a communication system. We are constantly reminded of transportation traffic because it is highly visible, especially when a traffic jam delays arrival at our destination. Because telephone traffic is invisible, we are not aware of it unless our calls are *blocked*, thereby inconveniencing us.

Both highway traffic and telephone traffic occasionally experience traffic jams that slow down or completely block traffic flow. From the user's point of view, the ideal situation would be never to encounter a traffic jam. However, this would require designing the highways or communication facilities so large that they would always meet peak demand. Obviously, this would not be feasible. From a practical point of view, user demand must be balanced with economic considerations.

The football fan en route to a big game might be understanding—although impatient—when experiencing a traffic jam caused by cars converging on the stadium. The same driver would not be as understanding of a similar delay twice a day, five days a week when driving to and from work. To provide good service and still be cost effective, highways must be engineered to accommodate busy-hour traffic, not occasional peak traffic.

Similarly, telephone networks experience peak traffic overloads (traffic jams) on certain holidays and times of unusual occurrences such as storms, earthquakes, and other disasters. Telephone networks are engineered using the same basic principles as highway design. They must be able to accommodate normal busy-hour traffic rather than occasional peak loads.

The mathematical description of message flow in a communication network is called *teletraffic theory*, a branch of applied probability.

The traffic engineer uses teletraffic theory and engineering principles to design efficient communication systems. This chapter discusses traffic engineering for telecommunication systems.

Traffic Engineering Procedures

Traffic engineering is the science of designing facilities to meet user requirements. The objective of telecommunications traffic engineering is to specify the quantities and arrangements of telephone trunks and switching equipment required to handle user traffic.

Traffic Studies

To design a major telecommunication system to user needs, the needs must first be determined. This is done by conducting a traffic study, which is simply a count of calls classified by types, such as incoming calls, internal calls, local calls, WATS-line calls, and private-line calls. Call counts are generally obtained by using call recording devices; however, in very small systems they might be obtained by manual methods.

Call data is usually summarized by half-hour intervals. This data is to identify the busy hour, the two consecutive half-hour periods in which the largest number of calls occurred. The process is repeated for five days and the average computed. The resulting number is the *average busy-hour traffic count,* a very important factor in determining user requirements, since equipment and trunks are provided to handle busy-hour traffic. To be truly representative, it is important that the week chosen for the traffic study be a typical five-day week that does not include a holiday. Figure 11.1 illustrates a traffic study summarization.

The busy hour in this study is from 10:30 a.m. to 11:30 a.m. The average busy-hour count is 349 (165 + 184). Similar studies are prepared for other types of traffic such as internal calling, WATS calling, OCC calling, and DDD calling. Figure 11.2 shows the hourly distribution of daily calls (from Figure 11.1). This graph demonstrates that approximately 15 percent of the total day's calls took place during the busy hour (10:30 a.m. to 11:30 a.m.). Figure 11.3 shows a record of daily average calls by months over several years data that is used to predict daily average calls by months. This record is also used to predict daily average calls for future months. Similar records are prepared for other types of traffic, such as internal calling, WATS calling, OCC calling, and DDD calling.

The purpose of conducting a traffic study is to obtain data for use in predicting future requirements. As a rule, fairly accurate predictions

Figure 11.1
Sample Traffic Study of Incoming and Outgoing Calls

XYZ Company
Traffic Study

Period May 9–13, 1984 **Incoming and Outgoing Calls**

	Mon	Tue	Wed	Thur	Fri	5 Day Total	Daily Average
9–9:30	70	68	63	71	73	345	69
9:30–10	131	126	122	134	137	650	130
10–10:30	139	129	126	143	153	690	138
10:30–11*	168	159	148	172	178	825	165**
11–11:30*	187	174	175	185	199	920	184**
11:30–12	162	156	152	163	172	805	161
12–12:30	136	132	137	140	145	690	138
12:30–1	94	90	87	93	96	460	92
1–1:30	132	122	136	135	145	670	134
1:30–2	148	136	137	138	151	710	142
2–2:30	147	142	140	146	155	730	146
2:30–3	141	138	135	146	150	710	142
3–3:30	165	146	149	158	172	790	158
3:30–4	168	165	152	165	175	825	165
4–4:30	142	136	129	150	153	710	142
4:30–5	129	125	138	141	148	680	136
Total Day	2,259	2,144	2,126	2,280	2,401	11,210	2,242

* Busy hour
** Average busy-hour traffic count

can be made by taking a five-day count each month for several years; however, the more data available, the more accurate the prediction will be. Some businesses are also subject to seasonal fluctuations. Thus, it is important that each month of the year be represented. Monthly data listed for several years will identify *busy-season* traffic, just as half-hour data identifies busy-hour traffic.

Predicting Future Telephone Usage

To determine the amount of equipment required to provide good service, the traffic engineer must not only have data on the number of calls but also on the duration of the calls. This data is obtained by periodically conducting a *holding time study. Holding time* is the conversation time plus the time the equipment is engaged in establishing the connection. In other words, holding time is the total time that the telephone receiver is off the hook. The number of telephone calls multiplied by the average number of seconds per call (holding time) equals the number of seconds that the equipment is in use. For example, if the system handled 363 calls in an hour and the average call was 300 seconds, the equipment would be in use 108,900 seconds

Figure 11.2
Hourly Distribution of Daily Calls

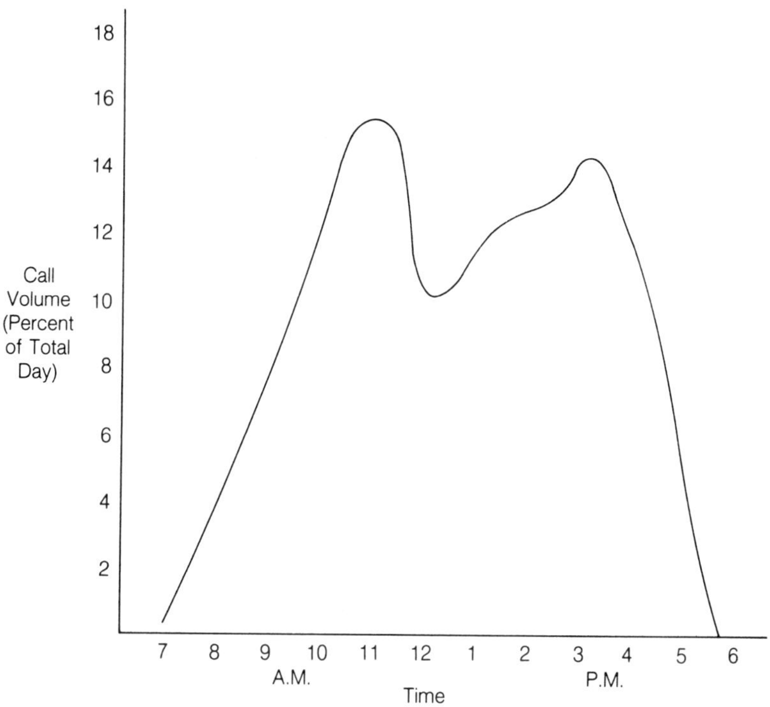

Figure 11.3
A Typical Record of Monthly Calls

XYZ Company
Record of Monthly Calls

Incoming and Outgoing Calls (Daily Average*)

	1981	1982	1983	1984	1985
January	1656	1754	1885	2210	
February	1675	1762	1912	2218	
March	1682	1776	1948	2232	
April	1690	1780	1976	2268	
May	1706	1796	1986	2242	
June	1710	1810	2022		
July	1650	1725	1850		
August	1625	1710	1810		
September	1750	1825	2129		
October	1726	1835	2142		
November	1781	1850	2168		
December	1795	1862	2185		

* 5 Day Average From Monthly Study

(363 × 300). This number is divided by 100 to determine the number of hundred call seconds (108,900 ÷ 100 = 1,089 CCS). *One hundred call seconds is known as 1 CCS.* (In the abbreviation, the first C is the roman numeral for 100, the second C stands for calls, and the S stands for seconds.) Additionally, any combination of calls and seconds totaling 100 would constitute 1 CCS; that is, two 50-second calls, five 20-second calls, or ten 10-second calls. Telephone usage is measured in terms of CCS.

The traffic engineer is charged with the responsibility of providing the quantities and configuration of equipment that will be required to handle a predicted volume of telephone usage. The data obtained from traffic studies is used to predict the future volumes of traffic usage. Since traffic loads tend to grow from year to year at varying rates, the traffic engineer studies the traffic data and selects a rate of growth that reflects past performance. The future call volume can then be predicted by multiplying present call volume by the anticipated growth rate and adding this predicted increase to the present call volume. Figure 11.4 shows a graph of monthly incoming and outgoing telephone calls used to predict future call levels.

Telephone systems are engineered to provide sufficient equipment to handle busy-hour traffic. Therefore, the percentage of the total day's traffic that occurs during the busy hour must be calculated. The number of busy-hour calls can then be predicted by multiplying the predicted future call volume by the percentage of calls occurring in the busy hour. The following example illustrates these calculations:

Present number of calls per day: 2,200 calls
Predicted growth rate (based on history): 10%
Predicted calls (2,200 × 10%) + 2,200 = 2,420 calls
Percent of calls in busy hour: 15%
Predicted busy-hour calls: 15% × 2,420 = 363 calls

In older telephone systems, CCS were computed manually by measuring holding time with stopwatch observations and multiplying the holding time figure by the number of calls. The resultant figure was changed to CCS by dividing by 100. To illustrate, if we counted 100 calls averaging 300 seconds per call, 300 CCS of usage would have been generated. (100 × 300 = 30,000; 30,000 ÷ 100 = 300.) Modern electronic switching systems provide direct readings in CCS, eliminating the need for this computation.

The CCS is one measure of usage. Another measure of usage is the *Erlang. One Erlang equals 36 CCS.* Since there are 3,600 seconds in an hour (60 × 60), 36 CCS (3,600 ÷ 100) equals one hour of

Figure 11.4
Average Daily Telephone Calls by Month

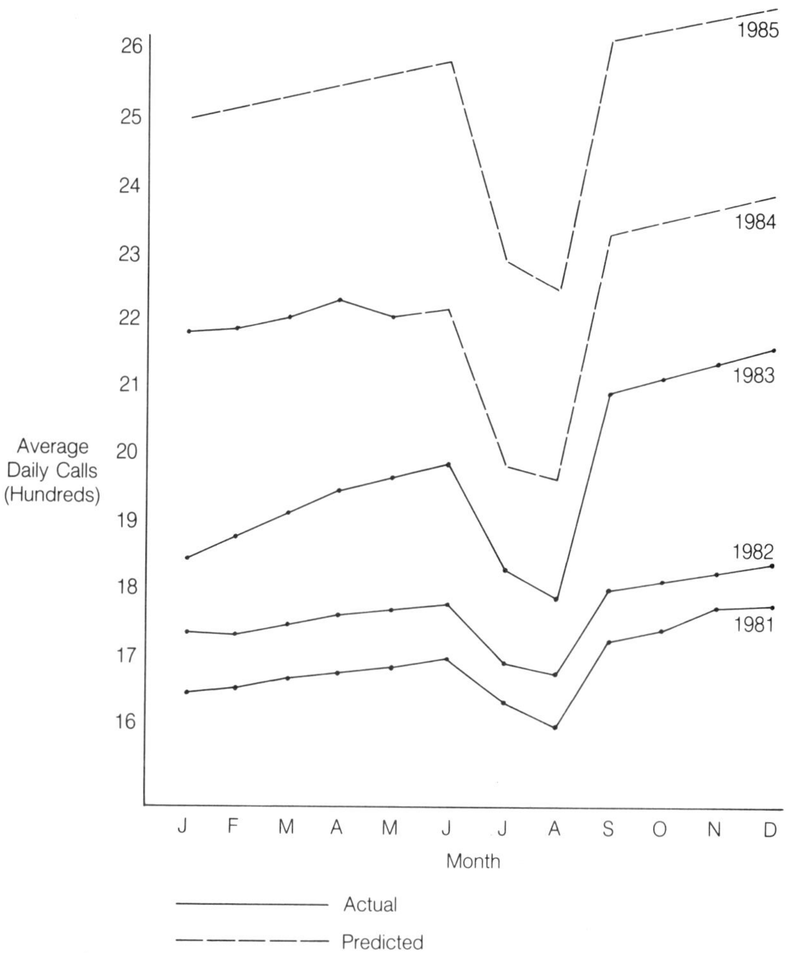

usage. Traffic engineers use both CCS and Erlangs in determining the quantities of equipment or trunks required. Figure 11.5 shows a sample conversion table between the two units of measurement.

The number of predicted busy-hour calls can be converted to CCS by multiplying the predicted number of busy-hour calls by the average holding time. Thus, if in our example the average holding time (based on history) was 300 seconds per call, this figure would be multiplied by 363 (the number of calls) and divided by 100, for a total of 1,089 CCS. To convert CCS to Erlangs, we divide by 36; i.e., 1,089 ÷ 36 = 30.25 Erlangs.

Figure 11.5
Conversion from CCS and Erlangs

Sample Conversion Table

Relationship between CCS and Erlangs

Call Seconds	CCS (Seconds/100)	Erlangs (CCS/36)
300	3	.083
1,200	12	.333
2,400	24	.666
3,600	36	1.000
10,000	100	2.777
18,000	180	5.000
36,000	360	10.000
72,000	720	20.000

Telephone Growth and Usage Prediction

The previous prediction of telephone usage was based on call volume data only. Another way to predict future usage is by using projections of anticipated telephone growth. The traffic engineer usually uses both methods, with one serving as a cross-check on the other.

In making projections, the traffic engineer needs to know the number of telephones to be served by the system, both at the time of system cutover and at the time the system reaches its maximum size. This information can be obtained either from the telecommunications manager or the Request for Proposal specifications.

The following example uses the number of telephones to be served by the system in future years to calculate the usage that will be generated by these telephones when the system reaches its maximum capacity. In this example it is assumed that the busy-hour calling rate and the holding time per call will not change over the life of the system. In actual practice, estimates of these two items for future years would be required, although calling rates and holding times change very little, if any, from year to year.

Present number of telephones	200	
Growth in number of telephones per year	20	
Telephones in service at the end of one year	220	
Telephones in service at the end of five years	300	
Daily calls—present (from traffic study)	2,200	(Figure 11.1)
Percent calls in busy hour	15%	
Busy-hour calls	330	(2,200 × 15%)
Busy-hour calls per telephone	1.65	(330 ÷ 200)
Average length of calls	300 seconds	

Usage is equal to the number of telephones times the number of calls per busy hour per telephone times the average length of telephone calls.

Usage at Cutover

$200 \times 1.65 \times 300 = 99{,}000$ call seconds
$99{,}000 \div 100 = 990$ CCS
$99{,}000 \div 3{,}600 = 27.5$ Erlangs

Usage One Year from Now

$220 \times 1.65 \times 300 = 108{,}900$ call seconds
$108{,}900 \div 100 = 1{,}089$ CCS
$108{,}900 \div 3{,}600 = 30.25$ Erlangs

Usage Five Years from Now

$300 \times 1.65 \times 300 = 148{,}500$ call seconds
$148{,}500 \div 100 = 1{,}485$ CCS
$148{,}500 \div 3{,}600 = 41.25$ Erlangs

Grade of Service

Every communication network is organized on the principle of sharing common equipment. When any shared equipment is provided in an insufficient quantity, there will be times when users will have to wait. Telephone systems could be designed so that users would never have to wait; however, such overprovision of equipment would be extremely costly and wasteful. Clearly, there must be a balance between the cost of providing service and the quality or grade of service provided.

A service delay occurs when a telephone number is dialed and a busy signal is received. The busy signal may occur for either of two reasons. First, the telephone line may be busy because a person is using it. Second, the dialed number is not busy but there are no available circuit facilities to reach the desired telephone. The second circumstance is referred to as a *network busy condition.* The term *blocking* describes a call that cannot be completed due to a *network busy condition.* When a call is blocked, the telephone user hears a faster busy signal than that received when the desired telephone line is busy. The standard signal to indicate a busy telephone line is a tone interrupted 60 times a minute (60 IPM); the standard signal to indicate a network busy condition is a tone interrupted 120 times a minute (120 IPM). The user is primarily concerned with getting the telephone call through and probably does not differentiate between these two types of busy signals. Similarly, the term *blocked* is used, slightly incorrectly, by the public to denote a call that cannot be completed because of either a line busy or network busy condition.

Grade of service is the probability of a call being blocked expressed as a percentage, such as P.05.

Cost Versus Grade of Service

Telephone facilities can be engineered to provide any desired grade of service. The lower the probability figure, the better the service will be and the more the system will cost. Management determines the organization's service objectives by balancing telephone system costs against the costs of inconvenience to users when their calls are blocked. The traffic engineer translates management's service objective into a specific grade of service quantification.

In the United States, operating telephone companies provide public telephone service engineered to P.01 grade of service. Private telephone systems are often engineered to provide a slightly poorer grade of service, generally in the P.01–P.05 range. Telephone users have become accustomed to the high-quality service provided by the United States public operating telephone companies (OTCs) and tend to use it as a frame of reference in judging all telephone service. Since users are particularly sensitive to changes in service rather than to absolute levels of service, any noticeable difference in service quality usually draws sharp criticism.

Traffic Capacity Tables

The traffic engineer uses traffic capacity tables to determine the quantities of trunks or equipment needed for a telecommunications system.The two most frequently used traffic capacity tables are the Poisson table and the Erlang B table. Both tables are based on mathematical probability theory and are named after the mathematicians who developed them.

Simeon Poisson was a nineteenth-century French mathematician who developed a mathematical model to predict the outcome of events. His theory was adapted for use in telephony in the early 1920s by a Bell Laboratories scientist named E. C. Molina. The Molina formula is used to predict the number of trunks required to handle various volumes of usage. Poisson theory and the Molina formula are based on the assumption that the sources of telephone traffic are infinite and that all unsuccessful call attempts (blocked calls) are retried within a relatively short time interval.

A. K. Erlang, a Danish engineer and mathematician, is often referred to as the father of teletraffic theory. The fundamental unit of traffic load, the Erlang, bears his name. In the second decade of this century, he developed a method of analysis used to predict the quantities of equipment required to handle a given volume of telephone traffic. The Erlang B traffic capacity tables are based on the assumption that the sources of telephone traffic are infinite but that all unsuccessful call attempts are abandoned.

Figure 11.6 shows a partial Poisson capacity table; Figure 11.7 illustrates a partial Erlang B capacity table. The tables have usage

Figure 11.6
Sample Poisson Capacity Table

Poisson Capacity Tables
Hundred Call Seconds at
Various Grade Levels

Trunks	Grade of Service At Indicated CCS Load			
	P.01	P.02	P.05	P.10
2	5.4	7.9	12.9	19.1
4	29.6	36.7	49.1	63.0
6	64.4	76.0	94.1	113.0
8	105.0	119.0	143.0	168.0
10	148.0	166.0	195.0	224.0
15	269.0	293.0	333.0	370.0
20	399.0	429.0	477.0	523.0
25	535.0	571.0	626.0	670.0
30	675.0	715.0	773.0	636.0
40	964.0	1012.0	1038.0	1157.0
50	1261.0	1317.0	1403.0	1482.0

To use this table find the usage in CCS in the appropriate grade of service column and read the number that appears on the same line in the trunks column on the extreme left.

Figure 11.7
Sample Erlang B Capacity Table

Erlang B Capacity Tables
Erlangs of Use at
Various Grade of Service Levels

Trunks	Grade of Service At Indicated Erlang Load			
	P.01	P.02	P.05	P.10
2	.153	.224	.382	.6
4	.870	1.093	1.525	2.0
6	1.909	2.276	2.961	3.8
8	3.128	3.627	4.543	5.6
10	4.462	5.084	6.216	7.5
15	8.108	9.010	10.63	12.5
20	12.03	13.18	15.25	17.60
25	16.13	17.51	19.99	22.80
30	20.34	21.93	24.80	28.10
40	29.01	31.00	34.60	38.80
50	37.90	40.25	44.53	49.60

To use this table find the usage in Erlangs in the appropriate grade of service column and read the number that appears on the same line in the trunks column on the extreme left.

(CCS or Erlangs) as one dimension and grade of service as the other. Usage quantities are aligned under each of the desired grades of service. To use the tables, locate the predicted usage load in the appropriate "grade of service" column and read the number that appears on the same line in the "trunks" column at the extreme left. For example, to use the Poisson table (Figure 11.6) for 105 CCS of predicted usage to provide P.01 grade of service, find the quantity 105 in the P.01 column and read the corresponding number in the "trunks" column. In this example, the number of trunks required would be 8. The Erlang B table is used in the same way, except that usage must be expressed in Erlangs.

Traffic Queuing

Traffic engineering is based on assumptions and probabilities, not certainties. The Poisson theory is based on the assumption that when a call is blocked, the caller will try again within a relatively short time. Erlang B theory assumes that the caller will not call again; that is, that the call is abandoned. There is no way to actually determine whether a caller will try again and, if so, when the retrial attempt will be made. There are so many variables that enter into the situation, such as the urgency of the call, time of day, and priority of other activities, that retrial attempts are impossible to predict accurately.

Telephone systems for some organizations are engineered so that many arriving calls experience some delay. These organizations include airlines, hotel reservation centers, service bureaus, and government agencies. Their systems are engineered on the assumption that the caller will be willing to wait, particularly if there is some assurance that calls are being processed in the order that they were received in the system. In these systems the cost of providing a better grade of service (less waiting time) is balanced against the likelihood of losing the calls by the inconvenience of waiting. Because of the nature of the calls to these organizations, it is assumed that callers will be almost certain to wait rather than abandon their calls.

Calls are said to be *queued* when they are waiting in the sequence in which they arrived in the telephone system. A queue is simply a waiting line. Customers are served when they arrive at the head of the line; newcomers enter the queue taking their place at the end of the line. Banks have refined the queuing procedure by combining many lines served by many tellers into a single line with access to the next available teller. This eliminates the possibility of the customer's selecting the wrong line and having to wait unduly long because of one customer's lengthy transaction. Supermarkets, on the other hand, generally require their customers to queue at each checkout station. Customers choose the queue in which they wish to wait, taking their chances on which line will provide the fastest service.

Most telephone systems do not have the capability to queue within the system. Call distributing systems, such as those used by reservation centers and government agencies and some least-cost routing systems do have queuing capabilities. In addition, distributed data processing systems are generally designed to provide queuing. Systems capable of queuing are generally engineered using a slightly different type of traffic capacity table, which is known as Erlang C or Crommelin tables. These tables add another dimension to the design process, since they take into account the number of calls to be held in queue as well as the grade of service and the predicted traffic load.

Telecommunication facilities are designed to transmit messages in either voice or data form. The principal differences between voice and data traffic are:

1. Data traffic originates from a finite, or limited, source; voice traffic originates from an infinite source.
2. Data traffic has a longer holding time.
3. Data callers will accept a poorer grade of service. Computers don't mind waiting; humans do and become impatient.
4. Data traffic is capable of being stored for later transmission and is adaptable to queuing techniques. Voice traffic includes two-way interaction; thus it is not generally adaptable to queuing techniques.

Telecommunication facilities are engineered to handle a predicted amount of circuit usage, regardless of whether the traffic is in voice or data form. Thus, the engineering principles are the same for both types of transmission.

The Decision-Making Process

The effective performance of a telecommunication system depends in large part upon the decisions made by the traffic engineer in the planning stages. If the system has been engineered to provide inadequate quantities of trunks and equipment, poor service is inevitable. If the system has been engineered for overprovision, the service will be good but excessively expensive.

After the traffic engineer determines the volumes of usage that the system will be required to handle, a decision can be made concerning the quantities of equipment and trunks to be provided. The traffic capacity tables (Poisson, Erlang B, and Erlang C) provide a guide to assist the engineer in making these decisions. The choice of table will depend upon the characteristics of the traffic to be served by the system. Figure 11.8 summarizes the characteristics of the leading capacity tables.

Figure 11.8
Characteristics of the Leading Capacity Tables

	Poisson	Erlang B	Erlang C
Usage Measurement	CCS	Erlangs	Erlangs
Traffic Source	Infinite	Infinite	Infinite/Finite
Blocked Calls	Prompt Retrial	No Retrial	Indefinite Wait
Type of Traffic	Random	Random	Random
Principal Users	OTCs	Private Systems and OTCs	Distributed Systems

Poisson, Erlang B, and Erlang C capacity tables were designed for public telephone systems with their corresponding large numbers of users. Therefore, none of the tables is perfectly suited for smaller private systems.

Communication theorists have developed other capacity tables designed specifically for use in private telephone systems. Dr. James Jewett of Telco Research Corporation, a private consulting organization, developed tables that he calls *Equivalent Queue Extended Erlang B (EQEEB)* to be used in the design of trunks that automatically route blocked calls to alternate routes.

The Center for Communications Management, Inc. (CCMI), another private communications consulting organization, has developed tables using a proprietary formula. These tables, known as *CCMI Pragmatist,* are based on the assumption that 70 percent of blocked calls will be retried during the busy-hour traffic period.

All traffic capacity tables are based upon statistical probability and their use is based upon certain assumptions. There are two fundamental factors that must be kept in mind when an engineer uses any traffic capacity table. The first is the mathematical truth that the combination of a precise number and less precise number or approximate number can only be as accurate as the least precise number. The use of capacity tables involves at least two dimensions. One dimension is an estimate of usage, an imprecise number. Another number is grade of service, a number which contains an assumed number of retrials, thereby making it imprecise. Therefore, regardless of the degree of precision built into the traffic capacity tables, the figures resulting from their use can only be as accurate as the least precise quantities and the assumptions made.

The second factor is that the accuracy of a statistical sample varies directly with the size of the sample; the larger the sample, the more accurate the prediction will be. Since private telephone systems are designed for a relatively small number of users (compared with the public telephone system), estimates of usage tend to be less accurate.

Thus, traffic capacity tables are a guide in directing the judgmental process; they are not absolute. The quantities of equipment specified in the capacity tables should be tempered with judgment, taking into account such other considerations as the firmness of the grade-of-service decision and the extent of the busy-hour level of traffic. Before reaching a decision, it is wise to consider the quantities of equipment suggested by several appropriate traffic capacity tables. In spite of the theoretical shortcomings of traffic capacity tables, they provide a valuable guide for decision making; their use contributes greatly to effective facilities engineering.

Returning to the previous example of usage for incoming and outgoing calls, consider how the traffic capacity tables are used as a guide in the decision-making process.

Since the traffic from the system under consideration is random, comes from an infinite source, and has no provision for queuing capabilities, the traffic engineer would use either Poisson or Erlang B traffic tables in determining the number of trunks required for incoming and outgoing calls. The following data summarizes the usage from the example and converts it into the number of trunks required at both P.01 and P.05 service levels using both Poisson and Erlang B capacity tables:

	Table	Trunks Required	
		P.01	P.05
Usage at Cutover			
990 CCS	Poisson	39	36
27.5 Erlangs	Erlang B	38	34
Usage One Year from Now			
1,089 CCS	Poisson	44	40
30.25 Erlangs	Erlang B	41	36
Usage Five Years from Now			
1,485 CCS	Poisson	57	53
41.25 Erlangs	Erlang B	54	47

In examining the differences between the numbers of trunks suggested by the Poisson and Erlang B tables, we see that although the two are very close, the Poisson estimate is slightly higher. The difference between the two tables is in handling blocked calls. The Poisson table assumes prompt retrial of blocked calls while the Erlang B table assumes no retrials. Since retrials are impossible to predict accurately, a possible decision might be to compromise halfway between the two figures.

In determining the number of trunks required, the traffic engineer would provide sufficient equipment to handle usage requirements for the life of the system (five years, in this example).

The trunks required to handle incoming and outgoing traffic are provided by the operating telephone company at a monthly rate. The engineer must provide enough equipment to terminate the ultimate number of trunks required, but the trunks do not need to be contracted for until they are needed. However, the ability to terminate the trunks must be foreseen and taken into account in the equipment provision.

A common fault in the engineering of private systems is to underestimate future requirements with the result that the system may not be able to meet user requirements over its life expectancy. This results in poor service and unnecessarily high replacement costs. It is generally more advantageous to err on the high side of the prediction, since this merely postpones system replacement.

System Capacity Evaluation

After a new telephone system has been installed, the manager will want to know whether the system has been properly engineered to provide the desired grade of service. To do this, the manager conducts a *load/service relationship analysis,* designed to test the system.

To conduct such an analysis, data is collected on the number of calls handled by the system and the number of blockages encountered. This data is obtained by reading registers provided within the system. The data is graphed to show the load/service relationship by plotting the call volume on the X-axis and percentage of calls blocked on the Y-axis, as illustrated in Figure 11.9. In a fully loaded system, the graph shows a definite bend, indicating a critical area where the percent blockage increases exponentially. Thus, the graph depicts the number of calls that can be handled before excessive blockage occurs, indicating that system capacity has been reached. In Figure 11.9 the critical area occurs in the 350–360 call range.

If the system is new and thus underloaded, there is a probability that no blockages will be encountered. In this event, the manager can force a system overload to test the equipment capacity, causing some trunks to be inoperative (falsely busy) to obtain traffic load/service data. This technique provides a reliable check on the adequacy of the trunks and equipment.

Summary

Telephone traffic is the flow of messages through a communication system. Traffic engineering is the science of designing facilities to meet user requirements.

User requirements are determined by conducting a traffic study, a count of telephone calls classified by types such as incoming calls, internal calls, local calls, WATS-line calls, private-line calls, and any other types of calls handled by the system. In addition to the data

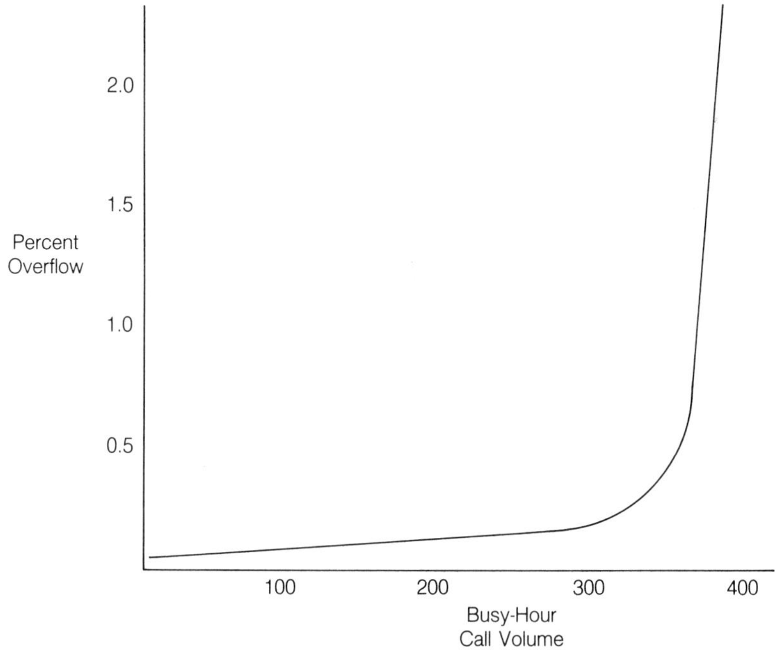

Figure 11.9
Load/Service Relationship of System Capacity

on the number of calls, it is necessary to collect data on the average length of calls (holding time).

Present call volumes obtained in the traffic study are used to predict future call volumes. Estimates of telephone growth are also taken into account in predicting future call volumes. Present holding time data is used to predict future holding times. The product of these two quantities is described as usage (call volume × holding time = usage). Usage may be expressed either in CCS or Erlangs.

A user receives a busy signal for one of two reasons: the line is busy because someone is using it, or there is no available circuit facility to reach the desired telephone. This second condition is referred to as a network busy condition. Grade of service refers to the probability of a call being blocked because of a network busy condition. It is expressed as a percentage. Thus, P.05 grade of service indicates that there is a 5 percent probability of receiving a busy signal because of a lack of circuit facilities during the busy hour.

Telephone facilities can be engineered to provide any desired grade of service. The lower the probability, the better the service will be and the more the system will cost. The traffic engineer balances the cost of providing service with the quality or grade of service to be provided.

The traffic engineer uses traffic capacity tables to determine the quantities of equipment required for the system. The two most frequently used traffic capacity tables are the Poisson table and the Erlang B table. Both tables are based on mathematical probability theory and contain certain assumptions. The Poisson theory assumes that when a call is blocked, the caller will try again within a relatively short time. Erlang B assumes that the caller will not try again; that is, the call will be abandoned. Traffic capacity tables have grade of service as one dimension and usage estimates as the other dimension. They are used as a guide in determining the number of trunks and/or units of equipment to be provided in the new system.

A new telephone system may be evaluated to determine whether it has been properly engineered to provide the desired grade of service by conducting a load/service relationship analysis. This analysis shows the grade of service the system provides at various call volume levels.

Review Questions

1. What is telecommunications traffic engineering, and what is its objective?
2. How is the average busy-hour traffic count computed? What is the purpose of obtaining this data?
3. Why is it important to identify busy-season traffic as well as busy-hour traffic? What are some types of businesses that experience seasonal fluctuations?
4. Why is it necessary to have holding-time data on calls?
5. If the average busy-hour call count was 312 calls and the average holding time 300 seconds, how many CCS would this represent? How many Erlangs?
6. If Jim calls Linda and receives a busy signal because Linda is talking to Dave, has Jim received a network busy condition? Explain your answer.
7. Which of the following figures represents the best grade of service: P.01, P.03, or P.05? Explain your answer.
8. Describe the condition of queuing. Which is more adaptable to queuing techniques—voice or data? Why?
9. What is the purpose of traffic capacity tables? Upon what factors does the choice of table depend?

References and Bibliography

Bishop, Lyman. "Systems Integration Hones Competitive Edge." *Communications News*, February 1984, 32–33.

Hoffman, Hugh. "Extended Erlang C: Traffic Engineering for Queuing With Overflow." *Business Communications Review*, Vol. 13, No. 4, July–August 1983, 28–33.

Jewett, James E., Jacqueline B. Shrago, and Bernard D. Yomotov. *Designing Optimal Voice Networks for Businesses, Government, and Telephone Companies.* Chicago: Telephony Publishing Corporation, 1980.

North American Telephone Association. *Industry Basics.* Washington, D.C.: North American Telephone Association, 1982.

Technical Staff, Bell Telephone Laboratories, Incorporated. *Engineering and Operations in the Bell System.* Murray Hill, N.J.: Bell Telephone Laboratories, Incorporated, 1977.

Williams, Philip. "Telephone Plant Operating System Mirrors CO Switching Process." *Telephony,* January 17, 1983, 44–48.

12 Theory of Rate Making

There are two times when peoples' attention focuses sharply on the telephone: when a telephone is inaccessible and when the cost of obtaining services changes dramatically.

Telephones have become so commonplace in our society that we take them for granted until something happens to remind us of their value. When our telephone is temporarily out of service, when we are away from our home or place of business and no telephone is available, or when service is denied temporarily for lack of telephone facilities, we are reminded of our dependence on the telephone—and impatient at not having instant service.

The other time we are reminded of the telephone is when we receive the monthy phone bill. As long as charges on the bill are reasonably in line with our expectations, we give the bill little thought. However, when a decided change in charges occurs, we immediately show interest in rates and pricing.

Fundamental changes are occurring within the telecommunications industry that recently have been reflected in our telephone bills and in our daily lives. Our telephone bills have become more complex, and we have to understand the new entries in order to interpret them. Many of us have begun to receive service from more than one telephone company and therefore receive two or more monthly telephone bills. Some of us have bought our telephones rather than continuing to rent them. Probably most of us have become more discriminating about when we call and over what telephone facilities. We all need education about obtaining telephone repairs. Overall, we probably have been overwhelmed by our options.

This chapter will examine the two principal theories of rate making and will describe the transformation of the telecommunications marketplace from one of regulation to one of competition.

Historical Perspective of Rate Making

When the Bell Telephone Company was organized in 1877, Alexander Graham Bell and his associate, Gardiner Hubbard, made a decision that proved to have far-reaching effects—the decision to rent telephone equipment while selling telephone service. This represented a departure from the accepted practice of public utilities, such as electric companies, which sold electricity but required their customers to own their own electrical equipment. As independent telephone companies entered the marketplace, they followed Bell's example and patterned their charging practices in the same way. This policy of requiring customers to rent their telephone equipment remained in effect for over a century.

The Communications Act of 1934

The Communications Act, passed by Congress in 1934, established the Federal Communications Commission and gave it the power to:

> regulate interstate and foreign commerce by wire and radio so as to make available, so far as possible, to all the people of the United States a rapid, efficient, nationwide and worldwide wire and radio communications service with adequate facilities at reasonable prices.[1]

This act resulted in the formulation of national policy regarding telephone rates, providing that:

1. telephones be priced within the reach of everyone
2. telephone companies be entitled to a fair return on their investment
3. all telephone companies post tariffs in advance and the state and federal commissions have authority to approve or suspend tariffs announced by the carriers
4. all telephone companies be permitted to interconnect to the long distance network and to share in long distance revenues in proportion to their investment and volume of business
5. state public utility commissions regulate intrastate telephone rates and the operation of intrastate activities of the telephone company

The Influence of Regulation on Rate Making

Three factors that have helped bring the price of telephones within the reach of everyone are:

1. cross-subsidization: local services subsidized by long distance services, rural services subsidized by urban services, and basic services subsidized by deluxe products

1. *The Communications Act of 1934,* Title I, Section 1, p. 1, 47 USC 151.

2. low depreciation rates: long life for telephone plant and equipment with accompanying long-term depreciation rates
3. low installation rates: installation charges capitalized; that is, charged to the investor rather than to the subscriber

Theory of Regulation

The theory underlying regulation is that regulation serves the public interest by protecting customers from the arbitrary exercise of monopolistic power. It assumes that regulated industries have special characteristics that make them a natural monopoly. Public utilities have been allowed to operate as regulated monopolies in order to prevent wasteful duplication of services. Regulation replaces market competition and is designed to protect the customer from excessively high service rates.

Tariffs Tariffs are the published rates, regulations, and descriptions governing the provision of communications services offered by a regulated utility. They are prepared by the utility company and submitted to the regulating commmission for approval. The commission studies the tariff and approves it when it is in accordance with the commission's policies relating to earning requirements and operating practices. All common carriers are regulated; each common carrier has its own tariffs to describe its service offerings and rates. There are no tariffs for interconnect vendors because they are not regulated.

Tariffs serve as a contract between the customer and the utility. They are written in language that is easy to understand. In the case of a disagreement, the commission is empowered to interpret the tariff and render a decision.

Both intrastate and interstate tariffs are required by law to be on file in the business office of the telephone company and available for public inspection. To properly control communications expenses, the telecommunications manager should be familiar with the tariffs affecting the services obtained from the telephone company.

Figure 12.1 shows a list of typical intrastate tariffs for an operating telephone company.

Figure 12.2 shows part of a typical exchange service tariff illustrating service offerings and monthly rates for basic service, telephone access charges, and telephone sets.

Figure 12.3 shows part of a typical message toll tariff.

Rate of Return The Communications Act provides that regulated telephone companies are entitled to a "fair return on their investment." The term *rate of return* refers to the ratio of the funds available for distribution to investors to the total funds invested. It should not

Figure 12.1
Typical Intrastate Tariffs

Intrastate Tariffs

1. Local Telephone Exchange Service
2. Auxiliary Services and Equipment
3. Message Toll and Assisted Call Services
5. Wide Area Service
6. Directory-Assistance Service
7. General Regulations
8. Private-Line Services
9. DATAPHONE Digital Service
10. Facilities for Other Common Carriers
19. Mobile Telephone Service

The absence of some numbers indicates that tariffs have been withdrawn. To avoid confusion, the number corresponding to a withdrawn tariff is not reused.

Figure 12.2
Typical Exchange Service Tariffs

Local Telephone Exchange Services

Exchange Service Offerings and Monthly Rates

EXCHANGE RATES					
Rate Groups:	A	B	C	D	E
Access Lines	1 to 20,000	20,000 to 50,000	50,000 to 200,000	200,000 to 500,000	Over 500,000
Business Services:					
PBX Trunks	$20.00	22.00	24.00	26.00	28.00
1 Party	9.50	9.50	9.50	9.50	9.50
Residence Services:					
1 Party Flat	$ 8.00	8.50	9.00	10.00	11.00
2 Party Flat	6.00	6.50	7.00	8.00	9.00
1 Party Measured	7.25	7.75	8.25	9.25	10.25
2 Party Measured	5.00	5.00	5.00	5.00	5.00

be confused with *telephone rates,* which are rates for telephone service.

In the telecommunications industry, the FCC and the state public utility commissions determine the net revenue requirements necessary to compensate the investors for use of their money. This rate must be high enough to attract investment capital and to preserve the financial stability of the utility. At the same time, the revenue requirement must be as low as possible to protect the telephone customers.

The FCC has jurisdiction over interstate revenues, which are derived principally from interstate long distance services. The state public service commissions have jurisdiction over intrastate revenues, which are derived from local area services and intrastate long distance services.

Figure 12.3
Typical Message Toll Tariff

Message Toll Service

Basic Rate Schedule—applies to each Message Toll Service call with certain discounts as specified.

Rate Step	Rate Miles	First Minute or Fraction	Each Additional Minute or Fraction
1	1–10	$ 0.08	$ 0.04
2	11–15	0.10	0.06
3	16–20	0.12	0.08
4	21–25	0.17	0.12
5	26–30	0.23	0.17
6	31–50	0.30	0.23
7	51–100	0.37	0.29
8	101–200	0.44	0.35
9	Over–200	0.50	0.40

Discounts from the Basic Rate:

A 30% discount applies on each call placed Monday through Friday and Sunday during the period from 5:00 P.M. to, but not including, 11:00 P.M.

A 30% discount applies on each call placed during the period from 8:00 A.M. to, but not including 5:00 P.M. certain holidays (New Year's Day, Independence Day, Labor Day, Thanksgiving Day, and Christmas Day), or their resulting holiday when said holidays fall on a weekday (Monday through Friday).

A 50% discount applies on each call placed during the period from 11:00 P.M. to, but not including, 8:00 A.M. Sunday through Friday, all day Saturday, and Sunday to, but not including, 5:00 P.M.

Additional Charges

Collect Call	$0.90 per call
Calling Card	.35 per call
Bill to Third Number	.90 per call
Request for Time and Charges	.90 per call
Person to Person	1.50 per call
Calling Card Person to Person	1.50 per call

The Value-of-Service Concept

Historically, telephone rates have been based on the value-of-service concept, which holds that rates for providing a service to a specific customer should be related to the value, or utility, of the service to the customer. Thus, the value of service for customers served by larger telephone exchanges should be greater than the value of service for customers served by smaller telephone exchanges because of the greater number of customers they can call on a local basis.

The revenue requirements of the telephone company are converted into telephone rates. Rates that employ the value-of-service concept are designed to promote widespread telephone availability and usage.

Size of Exchanges Telephone company exchanges are grouped according to the number of access lines they serve. For example, one group might include all exchanges serving up to 30,000 access lines. A second group might include from 30,001 to 40,000 access lines, while a third group might be comprised of all exchanges serving from 40,001 to 70,000 access lines. The local rates for customers in an exchange area serving a larger number of access lines are higher than those for an exchange area serving a smaller number of access lines. This reflects the value-of-service concept.

Business Telephone Service Similarly, the rates for business telephone service are higher than residential service rates within the same exchange area. Since a business cannot conduct its affairs without telephone service, and since the service is capable of producing revenue for the business, the value of the service is deemed to be greater for the business customer.

Individual-Line and Party-Line Service Many local exchange areas offer individual- and party-line telephone service. *Individual-line service* provides a customer with exclusive access to the serving central office, while *party-line service* shares access to the serving central office among two or more customers. (The term *individual line* should not be confused with *private line,* which refers to private facilities connecting two points for the exclusive use of one customer, such as tie lines.) Telephone rates are highest for individual-line service, and they decrease for service with larger numbers of parties. This is another application of the value-of-service concept. Clearly, the exclusive access to the serving office provided by individual-line service is of greater value than the limited access provided by party-line service.

Installation Service Historically, the cost of installing station equipment on customer premises has far exceeded the connection charge. This has helped to promote the expansion of telephone service.

Long Distance Services Traditionally, long distance charges have been based upon mileage, the distance from the point of origination to the point of termination. Both interstate and intrastate rates were designed to produce the amount of revenue that the respective commissions felt should be derived from long distance services. The mileage concept of rates is based on value of service, not costs of providing the service.

Since it would not be feasible to calculate mileage from one telephone to another, the country is divided into rate centers. A *rate center* is a geographically specified point used for determining mileage-dependent rates. The rate center of a telephone exchange is generally a point centrally located within the exchange area. All toll traffic originating within the exchange is assumed to originate from the rate center. The rate center is identified by locating its position in terms of vertical and horizontal (V–H) coordinates on a grid map. Similarly, terminating traffic is assumed to terminate at the distant rate center. The airline mileage between any two points is calculated by a mathematical formula using the V–H coordinates of both rate centers. The cost of each call can then be determined by use of a toll message rate table. These tables have mileage as one dimension and class of call (direct-dialed, person-to-person, collect, calling-card, etc.) per time interval as the other dimension.

The mileage method of computing toll charges averages the rates; it does not take into consideration the network facilities required to complete the calls. Thus, any two calls of the same type between any two equidistant points will cost the same.

Another method of pricing long distance telephone service is Wide Area Telephone Service (WATS), a pricing mechanism that offers a discount on telephone service to large-volume users. Under WATS pricing, the country is divided into a number of service areas, or bands, extending outward from, but not including, the customer's home state. A customer may subscribe to some or all of these service bands, and each service band added includes areas closer to the home state. WATS is available for a monthly charge based on the number of subscribed service bands plus a charge for usage (number of minutes of use to a particular service band). WATS is available for either inward or outward service; however, both are not available on the same line. If a customer desires both inward and outward WATS, it will be necessary to subscribe to both services.

Still another method of pricing long distance service is private-line service, in which the customer leases a telephone circuit, not interconnected with the public telephone network, for exclusive use. The private line may be used for transmission of voice, teletypewriter, data, television, or other signals. The line must be purchased for a specific purpose and is priced according to its transmission characteristics as well as the distance between the two connected points. Under regulation, the price of a private line is based upon average route costs.

Vertical services are services over and above what is required for basic communications capability; e.g., speakerphones, deluxe telephone sets, deluxe bells, or custom calling services (such as call waiting, conferencing, and automatic dialing).

When a vendor decides to offer a new telephone service, a price for the service must be established. Under regulation, the approach to pricing vertical services differs from the approaches for local exchange service and toll service. The pricing of vertical services is based upon the principle of optimum pricing. The optimum price is one that will yield an optimum contribution to earnings and thus permit basic services to be offered at prices lower than would otherwise be necessary to meet overall earning requirements. The optimum contribution is the largest contribution that is prudent, taking into account market conditions.

The pricing of vertical services is another application of the value-of-service concept. By virtue of their deluxe aspects, vertical services have a special appeal, and therefore value, to many customers.

In summary, the value-of-service concept, which was promoted by national policy and which had the approval of the regulatory commissions, resulted in some services being provided below cost and others being priced substantially higher than cost. These practices were deemed to be "in the public interest" and to promote the affordability of the telephone for everyone. The result was that residential services were subsidized by business services, local services were subsidized by long distance services, and basic services were subsidized by vertical services.

The Impact of Competition

The Carterfone decision permitted direct electrical connection of customer-provided equipment to telephone facilities, thereby opening the telecommunications marketplace to competition. Customers no longer had to rent telephones and associated equipment from the telephone company but could purchase this equipment from the interconnect vendor of their choice.

Tariffs for Interconnection

As a result of the Carterfone decision, tariffs prohibiting interconnection of private systems and equipment with the public telephone network had to be rewritten. The FCC had ruled that interconnections would be permitted as long as there was no adverse effect on the telephone company's operations or on the utility of the telephone network to others. The telephone company argued that unlimited interconnection of terminal equipment directly to the public network could cause damage to the public network. This argument was considered by the FCC, with the result that a registration program for all types of terminal equipment was established. Registration of equipment certified that it complied with technical specifications issued by the FCC. Interstate tariffs were revised to reflect the registration program.

The Competitive Market in Customer-Premises Equipment

Competition was first introduced in customer-premises equipment because the regulated carriers were not sufficiently innovative. Competition was expected to promote creativity in the design and manufacture of interconnect equipment, and it did. It also acted as a stimulus for cost-cutting techniques and, in general, has worked well. Given the freedom of choice, many customers—both business and residential—purchased their own telephone equipment. The competitive marketplace also stimulated the development of communications devices with the capability to interact with other electronic equipment, such as computers and word processors.

Specialized Common Carriers

Shortly after the Carterfone decision, the MCI case (1969) decided favorably on another aspect of competition—that of providing interstate long distance services. The decision permitted the MCI Corporation to provide long distance service based on costs over high-density telephone routes. The Bell System described the practice of competing in only the most profitable markets as *creamskimming* and contended that it was unfair. Bell argued that if the telephone company were to respond to creamskimming competition, it could result in the deaveraging of telephone rates and hinder the realization of universal telephone service. Uniform rates for both rural and urban subscribers would no longer be possible; rural subscribers would have to be charged more and urban subscribers less. The arguments of creamskimming were examined by the FCC and rejected.

As a result of these two decisions, regulated carriers with rates based on value of service had to compete with carriers whose rates were based on costs. Thus, regulated companies that subsidized local services by higher long distance rates and basic services by vertical services were at a disadvantage.

Rates Based on Costs

In a free enterprise economy, the price of a product or service is determined by adding a profit margin to the costs incurred in producing the product or service. Nonregulated entrants in the telecommunications marketplace base their rates and charges on the costs of providing the product or service.

Two major decisions caused a restructuring of telecommunications pricing policies: the Computer Inquiry II decision (1981) and the Modified Final Judgment (1982), which settled the United States Department of Justice antitrust suit of 1974 against AT&T.

Deregulation The Computer Inquiry II decision provided that customer premises equipment supplied by the Bell System be deregulated. The Modified Final Judgment stipulated that the CPE segment of the business be transferred to an unregulated Bell subsidiary. Subsequently, AT&T organized the American Bell Company to market and service customer-premises equipment in competition with other vendors. These events necessitated Bell's *unbundling* local service offerings by pricing their rates for switching services separately from prices for customer-premises equipment.

Since customer-premises equipment was no longer subsidized by other sources of revenue, it was inevitable that AT&T Information Services restructure its prices, basing them on costs.

Network Access The MFJ stipulated that the Bell System practice of sharing long distance profits with Bell and independent telephone companies must be discontinued. To compensate the operating telephone companies for their costs, the FCC directed that the unbundled rate for each business and residence telephone line be increased to cover the loss of the subsidy.

The FCC also directed that all long distance companies, including AT&T Communications, be required to pay the operating telephone companies long distance access fees. In return for these fees, the operating telephone companies were to provide equal access service to any long distance telephone company.

Local Access and Transport Areas The Modified Final Judgment ordered AT&T to divest itself of all its operating telephone companies. As a result of this divestiture, the spun-off BOCs were formed into seven regions. Their charter provided for monopoly carriage of local telephone traffic and its switching. Local calling areas were mapped into 160 *Local Access and Transport Areas (LATAs)* throughout the United States. The operating companies were empowered to handle intra-LATA calls and to charge all long distance carriers, including AT&T Communications (formerly Long Lines) for connecting calls to and from their LATAs.

Only long distance companies were empowered to provide telephone service between LATAs. No long distance company was required to serve any or all specific LATAs. Thus, if the costs of serving an area become so high that it is unprofitable, long distance companies are not required to offer service.

These provisions resulted in two very important consequences:

1. Long distance services no longer subsidize local telephone services.
2. No long distance company—Bell or other—is required to serve high-cost, low-density telephone routes.

An Industry in Transition

Telecommunications is a dynamic, technology-driven industry. Prior to deregulation, operating telephone companies had moved in the direction of cost-based pricing by introducing *measured service*, a method of pricing that is *usage sensitive*; that is, based on the amount of usage. In many areas, measured service replaced *flat rate service*, in which a subscriber paid a fixed monthly charge for unlimited telephone service. WATS, too, had undergone a change in pricing. Originally offered as a flat rate service providing an unlimited number of calls (thereby giving the impression that WATS calls were "free" if a WATS line were available), its pricing was changed to include charges for usage.

In an effort to reduce labor costs, telephone companies had promoted direct-dialing of long distance calls at reduced rates. To motivate more uniform traffic loading and hence use the telephone network more efficiently, the company had introduced discount evening, night, and weekend rates.

The year 1983 marked the beginning of deregulation in the telecommunications industry, the first step in what appears to be systematic, comprehensive regulatory reform. The divestiture of AT&T, with the division of responsibility among AT&T Information Systems, AT&T Communications, and the Bell operating companies, meant that users could no longer depend on "The Telephone Company" for everything. Users had to assume more responsibility for managing their own telephone systems.

Over the years, United States telephone users have experienced the best service in the world at comparatively low prices. The goal of universal telephone service became a reality. The rate policies of the regulatory commissions, along with technological advancements, protected telephone subscribers from the rampant inflation that pervaded most of the economy. In terms of constant dollars, telephone subscribers came out way ahead.

Recent changes in the FCC philosophy from one of protective regulation to one of market pricing have charted a different course for telecommunication rates. The termination of cross-subsidies as a result of the Modified Final Judgment has caused local telephone rates to rise. Countervailing influences have been provided by advances in technology and competitive long distance pricing in a market-driven economy.

Summary

The Communications Act of 1934 established national policy regarding telephone rates. A basic principle of this policy was that telephones be priced within the reach of everyone. Three factors that helped bring telephones within the reach of everyone were:

1. subsidization of local telephone services by long distance services
2. long life for telephone plant and equipment, resulting in low depreciation rates
3. installation services subsidized by investors

The theory of regulation is that the public interest is served by regulating agencies that protect customers from the arbitrary exercise of monopolistic power. Public utilities have been allowed to operate as regulated monopolies to prevent wasteful duplication of services. The published rates, regulations, and descriptions governing the provision of communications services offered by a regulated utility are known as tariffs.

Historically, telephone rates have been based on the value-of-service concept, which holds that rates for providing a service be related to the value of the service to the customer. Thus, rates for business telephone service are higher than the rates for residential service within the same exchange area, because the value of the service is deemed to be greater for the business customer. Similarly, the value of urban services was deemed to be greater that the value of rural services because of the greater number of telephones subscribers could call on a local basis.

The telecommunications industry is presently in a state of transition. After operating for many years as a regulated monopoly, the industry now finds itself partially deregulated and in a highly competitive environment. Nonregulated entrants in the telecommunications marketplace base their rates on the cost of providing the products or service, with no subsidization of one service by another. Recent FCC rulings and court decisions have made it necessary for regulated telecommunications carriers to eliminate subsidies for equipment and local service. To compensate for loss of cross-subsidization, regulated carriers have found it necessary to increase local rates. Countervailing influences have been provided by advances in technology and competitive long distance pricing.

Review Questions

1. What was the basic premise of the Communications Act of 1934?
2. What are the factors that have helped bring telephone service within the financial reach of nearly everyone?
3. What is the underlying theory of regulation? Why have public utilities traditionally been allowed to operate as regulated monopolies?
4. Explain the value-of-service concept and its role in telephone rate making.

5. What are vertical services, and how are they priced?
6. What two decisions opened the telecommunications marketplace to competition, and what two aspects of telecommunications were involved?
7. Briefly describe the changes in the telecommunications industry resulting from the Computer Inquiry II decision and the Modified Final Judgment.

References and Bibliography

Fogarty, Joseph D. "Capital Recovery: A Crisis for Telephone Companies, a Dilemma for Regulators." *Public Utilities Fortnightly,* December 8, 1983, 13–18.

Fowler, Mark. "We're Heading Ultimately Toward a Regulation-Free Telecom Market." *Communications News,* March 1983, 100.

Fritz, Jerald N. "Post Divestiture Terminal Equipment." Address before the National Association of Regulatory Commissioners, Detroit, Michigan, November 15, 1983.

Garfinkel, Lawrence. "Interexchange Telecommunications Markets in Transition." *Public Utilities Fortnightly,* July 21, 1983, 26–33.

Helling, Henrik. "Data Transmission Service Tariffs." *Telecommunication Journal,* Vol. 50, No. 4 (1983), 195–201.

Johnson, Leland L. "Why Local Rates are Rising." *Regulation,* July–August 1983, 31–36.

Levy, Robert. "Long Distance Phone Fight." *Dun's Business Month,* November 1982, 60–65.

Renshaw, Edward F. "On the Distribution of Telephone Communication Subsidiaries." *Public Utilities Quarterly,* July 21, 1983, 34–39.

Wilson, John D. "Telephone Access Costs and Rates." *Public Utilities Fortnightly,* September 15, 1983, 18–25.

Recent Developments in Telecommunications

13

The 1980s will be remembered as the decade of telecommunications—a decade that marked the changing character of the telecommunications industry. The deregulation of customer-premises equipment, the merging of data processing and telecommunications services, and the breakup of AT&T ushered in a period of confusion and momentous change. The new environment created opportunities for new suppliers, new products, and new services. The result was a dramatic increase in the number of vendors, a host of innovative products and services at alternative prices, highly competitive telecommunications marketing, and intense competition in long distance services.

The Technology

The combination of telecommunications and computer technology has been one of the most exciting developments of our times. The concept of distributed data processing depends upon telecommunications facilities to transfer data from one computer to another. The communications industry uses computer technology to switch voice and data messages over telecommunication lines and to record the details of long distance communications. Thus, each industry capitalizes upon the capabilities of the other for their mutual benefit.

The Development of Microelectronics

Electronics hardware has gone through three stages of development—from vacuum tubes to transistors to integrated circuits. (See Figure 13.1.) Each of these stages represented significant technological and practical advances. Among these were:

Figure 13.1
Point Contact Transistor

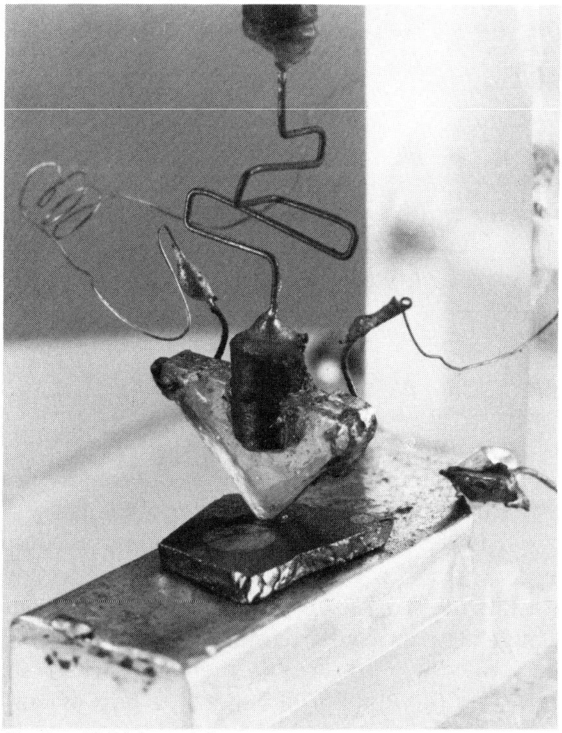

(Courtesy of Bell Laboratories)

1. a tremendous decrease in the size of equipment units
2. the capability to perform additional communication and computing functions
3. substantial decreases in production costs
4. reduction of problems caused by excessive heat
5. lowered power requirements
6. increased reliability
7. lower maintenance costs

With advances in electronic technology, computer components became smaller and smaller, reducing the amount of space required for circuit wiring and making it possible to combine several circuits on one circuit board. Later, many individual circuits were integrated into a single, complete electrical circuit on a small silicon chip. As the miniaturization trend continued, many thousands of circuits were integrated onto a single tiny chip, a process known as *large scale integration (LSI).*

Microelectric circuits are produced by a process similar to photolithography. The circuits are designed according to specifications, photographed, and reduced in size. They are then etched on a thin

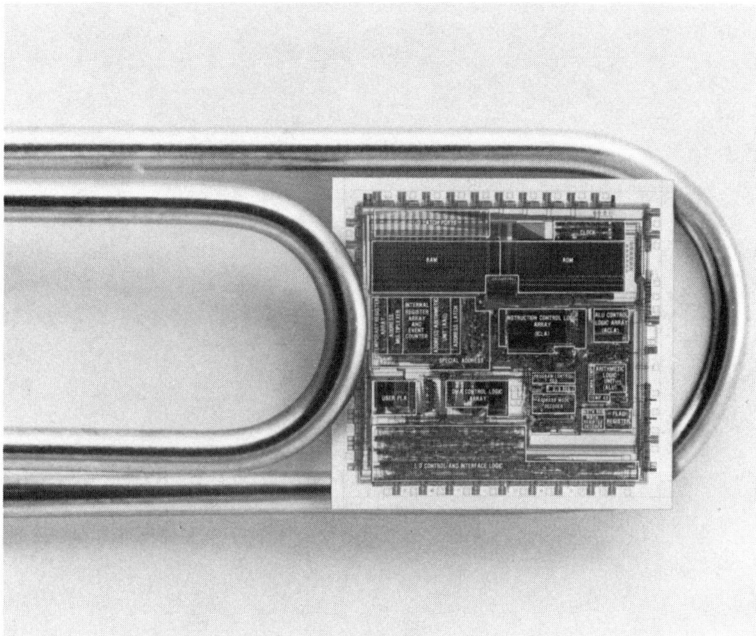

(Courtesy of Bell Laboratories)

Figure 13.2
The One-Chip Computer: Offspring of the Transistor

silicon wafer, a process that enables them to be mass-produced very inexpensively.

When the electronic circuitry for an entire central processing unit (CPU) is placed on a single, very small silicon chip, the result is a microprocessor (Figure 13.2). A *microprocessor* consists of a CPU, memory circuits, and input/output devices.

Microcomputers can function independently as standalone computers or can be linked into a computer or communications network. When microprocessors are incorporated into computer terminals, the terminals become intelligent, with the ability to perform certain processing functions independently; that is, without the power of the larger computer. Similarly, by incorporating microprocessors into telephones, the telephones become smart, with the ability to perform such processing functions as speed calling, automatic call back, hands-free operation, automatic answering, and call timing.

Advances in microelectric technology have increased both the number of components that can be placed on a chip and the speed and reliability of operation (Figures 13.3 and 13.4). An early industry standard was the 1K (1,000) chip that contained 1,024 memory cells on a single tiny chip. Each memory cell contained a storage transistor, or capacitor, which held an electric charge, and a switch transistor, which acted as a gate to permit the charge to enter or leave the

Figure 13.3
Digital Signal Processor
The digital signal processor chip is used in such applications as speech synthesis, voice recognition, filtering, tone detection, and line balancing in digital communication systems.

(Courtesy of Bell Laboratories)

Figure 13.4
Digital Signal Processor
The digital signal processor chip, slightly smaller than a telephone pushbutton, contains some 45,000 transistors and can make over a million calculations per second.

(Courtesy of Bell Laboratories)

cell. Present-day chips are predominantly 256K and contain 262,144 memory cells (256 × 1,024) on a silicon chip approximately ¼-inch square. A single 256K chip can store 10,000 numbers or 5,200 words of text.

The cost of microprocessor chips has also decreased substantially. (See Figures 13.5–13.8.) Chips that cost thousands of dollars a few years ago cost only a few dollars apiece today. Additionally, today's microcomputers have virtually as much computing power as medium-sized mainframe computers had a few years ago.

Barriers to Practical Applications

The advances in microelectronics have outdistanced our ability to put them to practical uses. This phenomenon is not new; it has been true of nearly all discoveries throughout history. The internal combustion engine was available for over fifty years before it was put to practical use in the automobile. Gasoline was a waste product in the oil-refining process until it was used to fuel automobiles. Television technology was demonstrated some 25 years before it became available commercially.

There are several factors that tend to delay the implementation of new technology. They include:

- ☐ absence of creative ideas for practical use
- ☐ economic considerations
- ☐ legal and political deterrents
- ☐ attitudinal barriers

Absence of Creative Ideas for Practical Uses Technological discoveries often occur as an offshoot of other research. Uses for these unexpected peripheral findings are not immediately apparent. Sometimes these types of discoveries are more important than the product of the basic research; however, the development of innovative ideas for their practical use depends upon the expensive trial-and-error process. Creative ideas cannot be developed upon demand; thus, there is usually a lag between any technological discovery and its practical application.

Economic Considerations Economic considerations play an important role in finding and developing practical applications for technological discoveries. Research activities are expensive and time-consuming. The high cost of implementing new technology must be weighed against the potential advantages they afford. Even when the cost-benefit analysis is favorable, the availability of investment capital can be a controlling factor.

Figure 13.5
Microprocessor BELLMAC 32

(Courtesy of Bell Laboratories)

Figure 13.6
Microprocessor BELLMAC 32A
The dime-size 32-bit chip contains almost 150,000 transistors and offers processing power comparable to that of today's minicomputers.

(Courtesy of Bell Laboratories)

(Courtesy of Bell Laboratories)

Figure 13.7
Microprocessor Chip
This microprocessor chip contains nearly 600 tiny, fast switches that operate at temperatures hundreds of degrees below zero. It is a step toward the day when high-speed computers, only about the size of a baseball, will handle the same amount of information now handled by a room-size computer, and do it faster.

(Courtesy of Bell Laboratories)

Figure 13.8
Microprocessor Chip

In the case of chip technology, the initial development was very expensive. Yet once the circuitry had been designed, the chips could be mass-produced and sold in large quantities. The costs of research, development, and tooling had to be recovered before the chip could be replaced with newer technology. In other words, product obsolescence had to be balanced against economic considerations. At one time the chip industry had made twofold capacity increases standard, but the high cost of retooling to manufacture a new chip ruled out anything less than a fourfold jump. This is demonstrated by the recent increase in chip capacity from 64K to 256K rather than to 128K.

In the telecommunications industry, many billions of dollars have been invested in analog transmission facilities. These facilities form the backbone of our communication system. They meet exacting maintenance standards and provide excellent communication service.

A newer type of transmission, digital transmission, has several important advantages over analog transmission. Digital transmission is ideally suited for data because there is no need to convert digital signals to analog signals for transmission. Additionally, noise and distortion are practically nonexistent in digital systems.

If the telecommunications industry were starting all over today, it would build all digital facilities. However, the staggering amounts of money that would be required to convert the entire system mandate the retention of the existing analog facilities except for system growth or replacement of obsolete equipment. The conversion from analog to digital facilities will come about gradually. Although digital technologies such as satellites, fiber optics, and carrier systems are rapidly being provided in new facilities, it will be a long time before existing analog systems will be replaced. Similarly, in spite of the superiority and widespread availability of *electronic switching systems (ESS),* millions of telephones will probably continue to be served by electromechanical switching systems for some time.

Legal and Political Deterrents In a regulated industry such as telecommunications, legal and political forces are major determinants of industry policy. National policy was established by Congress in the Communications Act of 1934. Although the technological environment has undergone sweeping changes over the last fifty years, the legal environment has remained relatively constant. Regulatory agencies must work within a legal framework in interpreting the Communications Act. They prescribe the rates of return on invested capital, rates of depreciation, and methods of raising capital for telecommunication companies. This environment has done little to promote the introduction of new products and services.

Many years ago the rate of depreciation for telecommunications equipment was established on the basis of a 20-year life expectancy. Once this depreciation rate was adopted, the industry was unable to retire equipment before its investment had been recovered through depreciation. Thus, regardless of technological breakthroughs, telecommunications companies were unable to use them until older equipment was fully depreciated. Regulation that prescribes long life expectancy for telecommunications equipment deters technological advances.

Attitudinal Barriers The implementation of new technology involves change—in procedures, equipment, required skills and knowledge, organizational relationships, and/or work environments. People generally fear change; it disrupts the comfortable status quo and evokes anxiety about their ability to master the new technology. Resistance to change is a natural phenomenon.

A study of the uses of telecommunications by Dordick and his colleagues (1981) concluded that "existing attitudes about communications technologies were one of the most complex barriers to surmount in getting people to use the new telecommunications techniques."[1]

The New Era in Telecommunications

The current telecommunications revolution is increasingly affecting our daily lives. Our homes and offices are rapidly becoming an electronic environment. New telecommunications services are changing our work patterns, leisure activities, transportation, education, manufacturing, health care, news media, and government.

The most dramatic development in telecommunications has been the combination of computers and communications services for home use. Personal computers allow the home to be connected by telephone lines to a centralized data base, permitting home users to access and interact with many different types of information services.

A recent Bell System advertisement stated:

> At this very moment, some homeowners are using the telephone network to dial up news and weather maps on their home video screens. They can shop from department store catalogs. And even compare supermarket specials.

1. H. S. Dordick, H. G. Bradley, and B. Nanus, *The Emerging Network Marketplace* (Norwood, NJ: Ablex, 1981), cited by Frederick Williams, *The Communications Revolution* (Beverly Hills, Calif.: Sage Publications, 1982), p. 178.

> With their ordinary home telephone lines providing transmission to a video screen, they can dial up information on seats on airlines or in local restaurants. They can even bank at home.
>
> Soon, local utilities will be able to use telephone lines to take remote readings of your gas and electric meters. And you'll be able to use those same telephone lines to monitor and remotely control home energy usage.

The following section will describe some of the applications of new telecommunications technologies. Nearly all of the services and products described are presently available.

Services for the Home

Videotex Services *Videotex* is a generic term used to describe a new group of consumer-oriented electronic information services that enable users to retrieve information stored in a central computer on a television or microprocessor monitor screen. Videotex includes both *teletext,* a one-way transmission system in which data signals are transmitted over the FM portion of a television signal, and *videotext,* a two-way, interactive system that is transmitted over telephone lines. Videotex systems combine television, telephone, and computer technologies.

The concept of videotex originated in the late 1970s; early development and trials took place principally in Europe. In England, the generic term is *viewdata.* The world's first public videotex service, *Prestel,* was formally launched by *British Telecom* in 1980, after a public trial lasting several years. Systems launched or under development in other countries include *Blidschirmtext* in West Germany, *Viditel* in Holland, *Teletel* in France, *Telidon* in Canada, and *Captains* in Japan.

In those countries the government exercises much greater control over the telecommunications system than in the United States. With the government paying the bills, the systems can be implemented with no concern about their ability to generate a profit. In this country, the government has not taken an active role in developing videotex, and the United States has lagged behind other countries. Since no version of videotex has been designated by the FCC as the official national standard, development proceeds cautiously.

The use of videotex service requires either a videotex terminal or a microprocessor. A videotex terminal consists of a control console connected to a television set and to a conventional telephone line. A microprocessor requires connection to a telephone line by means of a modem. The user accesses the service by dialing a local telephone number that connects the terminal to a computer system containing information and programs. The user can select specific files, and the desired data will be displayed on the screen of the television set or the microprocessor.

There are generally three charges for videotex service: a subscription fee, the telephone call (usually at local call rates), and a charge for the time the system is used.

Information Utilities The term *information utility* describes a system that supplies information just as traditional utilities supply water, gas, electricity, or telephone services. Information utility systems connect many small computers to a large central computer via telephone lines. The central computer can receive and store messages in electronic "mailboxes" where they can be retrieved by the persons to whom they are addressed. The central computer also contains data bases that users access for news reports, stock quotations, electronic mail, educational programs, sports reports, weather bulletins, and shopping services; users view the data on their own display screens.

The oldest videotex service in the United States is the Dow Jones News Retrieval Service. This service has the largest subscriber base of any online data base system. It provides current stock market quotes, detailed corporate information, economic reports, and a text search service. Text search permits a user to retrieve business articles from the *Wall Street Journal* and other business periodicals by category or keyword.

Many other data base services are currently available. Two popular services are CompuServe and The Source, both of which cater to microprocessor users. The CompuServe Information Service offers comprehensive news, sports, and weather reports (electronic newspapers), stock market updates, home shopping services, educational reference services, and electronic mail service.

There are many applications of the videotex concept, including home banking services, teleshopping, entertainment information services, libraries, tele-education, telemedicine, telecommuting, electronic mail, and the electronic office.

Telemedicine Telecommunications technology makes it possible to provide health care services from a distance, a concept known as *telemedicine*. Under this concept a remote location is linked by two-way television and audio signals with a hospital or other health-care facility. Physician and patient can interact directly. The equipment enables doctors to conduct patient examinations, inspect X rays and electrocardiograms, study microscopic specimens, diagnose medical conditions, and prescribe treatment. The technology also permits doctors to access medical reference sources and to consult with medical specialists.

Since there are many remote areas where doctors are in short supply and since medical equipment is extremely costly, telemedicine has enormous potential for improving health care. It also makes more efficient use of medical personnel and equipment.

Electronic Funds Transfer Computers and telecommunications make possible the electronic transfer of funds between accounts from one bank to another, eliminating the need for paper transfer instruments such as checks or cash. Home banking services generally permit two functions: information and accounts management (balance inquiries and transfer of funds between accounts) and bill paying.

Teleshopping Home shopping allows the user to see a product or service advertised on the television or microprocessor screen, order it via terminal, and arrange for payment by credit card.

Entertainment Information Services These services provide information about restaurants, movies, plays, places to go, and transportation schedules. Additionally, it may also be possible to make reservations or purchase tickets for a specific activity through videotex terminals.

Education The concept of computer-assisted instruction (CAI) has interested educators for some time, but until recently the high cost of computers deterred its implementation. With the advent of the microprocessors, inexpensive microcomputers have made CAI feasible for use in the classroom and for delivering instruction to the home. Classroom use of CAI offers students the opportunity for high-quality, impartial, individualized instruction. Computers can identify learning problems and adapt instruction to the specific needs and learning pace of the students. They impose discipline upon learners, forcing them to attain mastery levels of competence before progressing to the next step.

CAI designed for use in the classroom can also be used in remote locations, thereby permitting instruction to be delivered to the home. Some colleges have implemented *telecourses*, freeing students from commuting to and from the campus. The opportunities offered by telecommunication will certainly be instrumental in fostering lifelong learning.

Libraries The combination of computers and telecommunications is changing libraries into electronic information centers where users can access a central data base with a microcomputer. The data base may contain indexes, abstracts, citations, bibliographies, papers, or even full-text records. Computerized indexes permit users to search for items on a particular topic at a terminal screen; no card catalog is necessary. Computerized information systems can be accessed from anywhere in the world via telecommunications links, and many users can access the same data base at the same time. Homes equipped with videotex capabilities can also take advantage of the resources of electronic libraries.

One of the many data bases currently available online is MAGAZINE INDEX, an electronic *Reader's Guide to Periodical Literature.* When users key in topical words, this index searches for and displays relevant citations. Another online data base is Information Bank, a subsidiary of the *New York Times.* Information Bank is a general news data base of items abstracted from some 60 U.S. and foreign newspapers, magazines, and trade publications.

Electronic Polling Videotex technology, with its two-way, interactive characteristics, offers users the opportunity to voice their opinions about issues by participating in electronic polls.

A simplified forerunner of electronic polling took place in connection with the 1980 Reagan–Carter debate. The television network asked viewers to vote for the one whom they thought won the debate by calling a special telephone number. A different telephone number was assigned to each candidate. The voter indicated his or her choice by calling that candidate's number. Within a few hours, millions of calls were registered on an AT&T computer, and Reagan was declared the winner.

Videotex service permits the subscriber to make and register choices by following instructions displayed on the terminal screen. In teleshopping, for instance, viewers select the merchandise they want to purchase by pressing a key on the console to indicate their choice of the available options. This principle can be adapted for use in a variety of applications, such as conducting a survey or holding an election.

Electronic polling has the potential to stimulate interest in current affairs and make it easier for people to vote. However, its use is limited today because electronic polling requires the use of videotex-type equipment. Since the majority of voters do not have such equipment, participants in an electronic poll would not be a random sample of the general population. Therefore, the results could not be used for predictive purposes. Additionally, electronic polling would not be valid for general elections unless every qualified voter had access to an appropriate terminal.

Two-Way Communications The services previously described involve dialogue between a person at a computer and a data bank.

There are also other applications that involve a dialogue between two or more persons, each using a microprocessor (or other computer) linked to each other via telecommunication lines. Some of these applications are computer conferencing, electronic mail, and telecommuting.

Computer conferencing is a visual form of conference telephone call. The conferees talk to each other by keyboarding messages and transmitting them over telecommunications facilities to other computers connected to the system. Each conferee can provide his or her input at any time; there is no need to wait until another is finished transmitting. The computer stores all messages, permitting a conferee to leave the conference at any time and to retrieve messages when convenient.

Home computers can be programmed to provide message transmission capabilities to other computers over telecommunication lines. This service is known as (private) electronic mail and can be used 24 hours a day. Private electronic mail service has the advantage of nearly instantaneous delivery, in contrast to some commercial offerings that provide electronic transmission but delivery by courier or conventional mail service.

The increasing use of home computers permits two-way communication between home and business, providing a means for employees to work at home. Since this eliminates the need for traveling to and from work, it is often called *telecommuting*. Most types of white-collar or information-processing work lend themselves to this concept.

While the work-at-home concept is of obvious benefit to handicapped persons, parents, and others unable to commute, its benefits also extend to professional persons. By having a workstation in their homes, executives, lawyers, authors, and others who depend upon creativity in their work can extend their productivity. The flexibility of working at home allows these persons to "strike while the iron is hot" and capitalize upon their spontaneous ideas in an efficient, work-conducive environment. Offering workers the option of home employment can also contribute significantly to improved worker morale.

Business Services

Videotex Services Although the concept of videotex was oriented toward transmitting information electronically to homes, businesses also found uses for the technology. Some of the applications used by businesses include electronic funds transfer, news summaries, economic reports, market quotations, text search, and travel reser-

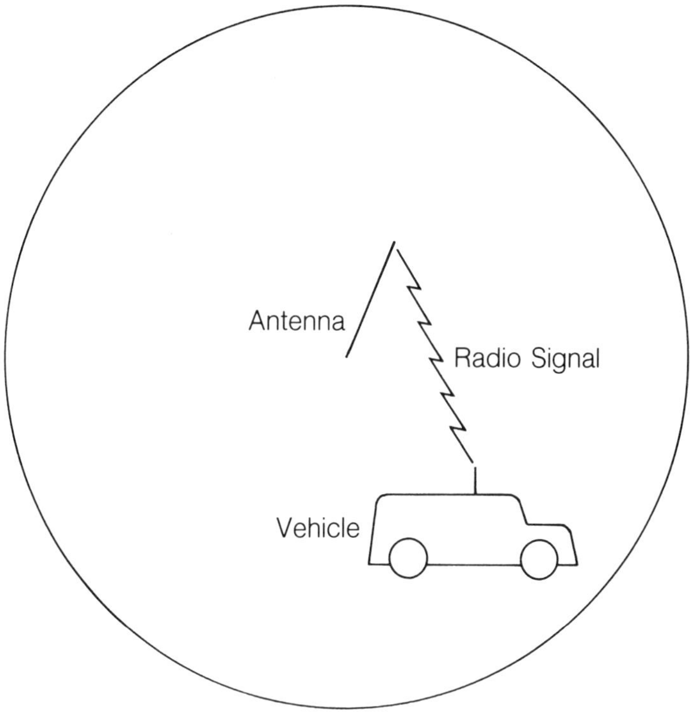

Figure 13.9
A Conventional Mobile Service Area

vations. In addition, business and industry use computer-assisted instruction for in-house training and development of their employees. Similarly, computer conferencing and electronic mail services are also widely used in business operations.

Cellular Radio *Cellular radio* is an advanced form of mobile telephone service. It combines radio and computer technology to provide telephone service to vehicles on the move.

Conventional mobile service uses two-way radio to connect the vehicle to the telephone system (Figure 13.9). The radio circuit establishes a talking path between the vehicle and an antenna connected to the telephone network. A single antenna serves an average-sized city, and communications can be maintained as long as the vehicle stays within range of the antenna. However, conventional mobile service has several drawbacks, including:

1. the limited number of radio frequencies available to provide service
2. inferior transmission characteristics provided by a single, distant antenna
3. interference from other vehicles on the same channel
4. high service cost

Figure 13.10
Cellular Radio Broadcast Areas

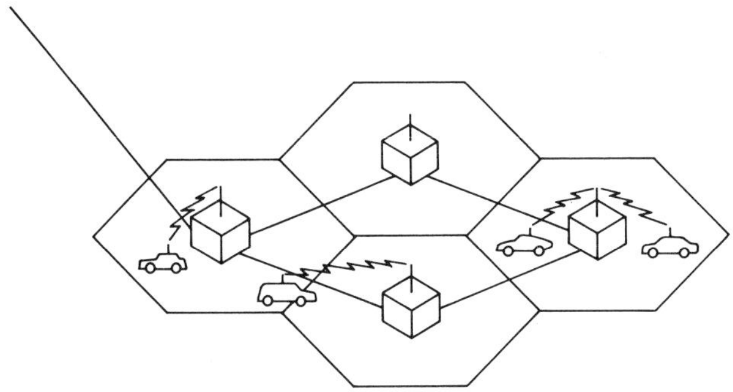

Despite these disadvantages, mobile service has been very popular. In fact, in many areas there is a very long wait to obtain service.

Cellular radio is a new concept designed to improve mobile telephone service and to make it available to more customers. The basic principle of cellular radio service is the division of one mobile telephone service area into a number of smaller areas called *cells* (Figure 13.10). Each cell contains a low-powered radio communicating system that is connected to the local telephone network. The individual systems replace the single high-powered system that formerly served an entire area and permit one radio frequency to be reused many times. A vehicle located in a cell area is connected to the telephone network through this low-powered radio system. When the vehicle leaves the cell area, radio contact is automatically transferred to the radio system of the adjoining cell as the vehicle enters.

The chief advantages of cellular radio are improved transmission and the ability to serve many more customers. Based on the long wait to obtain conventional mobile service and the spectacular demand for cordless telephones in the past few years, the demand for cellular radio should be heavy. Although the present price for mobile telephone service is relatively high, it is probable that market competition, mass-production techniques, and increased demand will ultimately result in lowered prices.

Voice Message Service *Voice message service* is an advanced form of telephone answering service that is integral to electronic telephone systems. It permits a caller to send a one-way spoken message to a service user. The message is stored in the designated *message box*, which is an assigned area in the system's computer. The area can only be accessed by the individual identification number given to each user. The message can be retrieved at the convenience of the recipient, allowing both sender and receiver to transact business at their own convenience.

This service eliminates concern over the possibility of missing important telephone calls. It also improves productivity by reducing wasteful attempts to reach an unavailable person.

Voice Processing The technology of computers speaking, storing human voices, and understanding and reacting to human speech is known as *voice processing*. Three types of voice processing systems are:

1. voice response (synthesized speech)
2. voice recognition
3. voice store and forward

Voice response is the conversion of computer output into spoken words and phrases that a human being can hear and understand. The computer combines various frequencies of electrical impulses stored in its memory to create vowels and syllables. Under the direction of a computer program, the vowels and syllables are synthesized into audible words, creating responses to program requests.

Voice recognition is the ability of a computer to understand and react to the human voice rather than only being able to accept typed commands. Voice recognition is a new technology still undergoing laboratory testing; it is limited by the computer's ability to understand human speech.

Today, computers with voice-recognition capabilities have limited vocabularies and are very expensive. As voice-recognition systems are further developed, their large-scale use is virtually assured, because it is easier to speak than to keyboard.

Voice store-and-forward systems enable a computer system to accept a message and store it until a transmission path is available or until the desired party calls to retrieve it. The advantage of store-and-forward systems is that messages aren't missed because the recipient is unavailable at the time the message is delivered.

Teleconferencing Teleconferencing allows groups of persons to assemble in two or more locations linked by telecommunication facilities to exchange ideas and information instead of meeting together in person. It reduces the costs of time and travel. Teleconferencing service has been available for many years; however, up until the last few years, the number of organizations using it was small.

The basis of all teleconferencing systems is audioconferencing. Based on recent sales figures collected by TeleSpan, a private consulting firm, the teleconferencing industry is now doubling in volume on an annual basis. Audioconferencing systems are growing at a faster rate than the more sophisticated videoconferencing systems.

The price differential between these two systems—which is substantial—probably accounts for the difference in growth rates.

Radio Pagers Another form of telephone service made possible by radio technology is paging service. Like mobile telephone service, radio pagers provide a way to communicate with someone on the move. The principal difference between mobile telephone service and radio paging service is that mobile telephone service provides two-way communication while radio paging is limited to one-way service.

Traditionally, radio pagers were used primarily by doctors and service personnel, who carried small units called *beepers* attached to their clothing. The radio transmitter broadcast a signal that could only be recognized by one unit to inform the user that a message was waiting. To receive the message, the user would go to a telephone and call a predetermined number. Because the signal transmission time was very short, many beeper units could be served by one transmitter. Recent technological developments have expanded the usefulness of radio pagers. Now, in addition to tone-only pagers (beepers), there are tone-and-voice pagers, pagers with digital displays, alphanumeric pagers, and pagers that print. Tone-and-voice pagers, as the name implies, emit both a beep and a brief voice message. Pagers with digital displays can receive digital information displayed on a tiny screen or printed out on a strip of paper. The information can be either all numbers or combinations of letters and numbers.

From a technological standpoint, nationwide paging is also possible. Nationwide paging will use telephone networks to route paging messages to radio transmitters located all over the country.

Paging service is growing rapidly, and its use is expected to expand dramatically as the cost of equipment continues to be reduced. Costs of paging units have dropped from $300 per unit to about $100 in the last few years, with the expectation that there will be further reductions to about the $50 level in the near future.

The Electronic Office

Many articles have appeared in newsmagazines and professional journals documenting the fact that the office environment is changing dramatically. A recent yearbook prepared under the direction of the National Business Education Association (NBEA) was entitled *The Changing Office Environment*. The change agent responsible for the new office environment was electronic technology; the reason was primarily economic.

The office was one of the last holdouts in embracing automation. During the 1970s, office productivity increased only 5 percent, while factory productivity rose 85 percent. With the demands for information increasing and workers' salaries rising, it was inevitable that organizations would seek ways to boost office productivity.

The goal of office automation is to speed up the flow of information while reducing operating costs and providing workers with interesting, challenging work experiences. The electronic office has been described variously as the *paperless office,* the *automated office,* and the *office of the future.* It is characterized by the automation of information-processing functions through the use of computer and telecommunications technology.

The office of the future has been the subject of much speculation; many persons have given us their impressions of what it would be like. Because of rapidly changing technology, no one knows what the office of the future will really be like; however, it will be characterized by the electronic storage, processing, transmission, and retrieval of information. A corollary will be the trend toward the decreasing use of paper.

Many of the components of the office of the future are used to a limited extent in organizations today. However, for the most part, the components are standalone processing units that function on an independent basis. Intelligent terminals have come into fairly widespread use; some are compatible, most are not. Centralized printers and corporate records cannot be accessed by standalone units. As a result, the same information is being duplicated by several departments instead of being available to each via a centralized data base. For example, a report prepared by the accounting department may contain data needed by the personnel department; the same material is typed by both departments. The principal difference between the office of today and the future office is that the various components will be integrated into a network configuration that permits each component to access a centralized data base.

The electronic office consists of a central information storage and computing unit, input and output terminals, and central printers, connected via telecommunication lines. The system can be shared easily by all users, updated constantly, and provide information in any useful form. The focal point of the system is the central information storage and computing unit that functions as a data bank and a data processor. The input terminals may be word processors or microcomputers. The availability of portable microcomputers has made it possible for office personnel to connect to the system wherever a telephone is available. Information may be retrieved on a terminal screen, by typewriter printout, by high-speed printers, or by computer output microfilm (COM) recorder. Terminal-to-terminal transmission, computer-to-copier transmission, and facsimile are all forms of electronic mail.

The total information system may consist of only a few components with limited capabilities, or it may consist of a wide variety of intelligent components with sophisticated capabilities. It may include voice processing, word and data processing, reprographics, records processing (electronic files), micrographics (film-stored records), facsimile, and, of course, telecommunications. The system is also capable of using videotex services provided by outside information service vendors. A security system using unique passwords protects the information and data from unauthorized access.

An electronic telephone system is also an integral part of the electronic office. The system can provide teleconferencing, either audio or video, and is equipped with smart electronic telephones with the service features required by the organization for the conduct of its business. The electronic office depends upon telecommunications links so that the various parts of the system can work together; thus, telecommunications makes the electronic office possible.

By instantly transporting information to wherever it is required, telecommunications permits managers to make informed decisions based upon timely, relevant information. By making information available to all departments and persons needing it and avoiding duplication, telecommunications can speed up operations, increase the amount of information handled, and increase productivity. For these reasons, computers and telecommunications are certain to be incorporated increasingly in business management information systems.

Summary

Advances in microelectronics have outdistanced our ability to put them to practical uses. Factors that delay the implementation of new technology include absence of creative ideas for practical uses, economic considerations, legal and political deterrents, and attitudinal barriers.

The current telecommunications revolution is increasingly affecting our daily lives. Our homes and offices are rapidly becoming an electronic environment. The most dramatic development in telecommunications has been the combination of computers and communications services for home use. Some of the new videotex services for the home include telemedicine, electronic funds transfer, teleshopping, entertainment information services, education, library services, electronic polling, computer conferencing, and private electronic mail. Home computers also permit telecommuting (working at home).

Businesses use such services as videotex, cellular radio, voice messaging, teleconferencing, radio paging, word and data processing, electronic mail, reprographics (facsimile), records processing (electronic files), and voice processing. The electronic office depends upon telecommunication links to enable the various parts of the system to work together.

By making information available to all departments needing it and avoiding duplication, telecommunications speeds up information flow and increases productivity. Thus, computers and telecommunications are certain to be incorporated increasingly in business management information systems.

Review Questions

1. How does the computer industry incorporate telecommunications technology in its operation?
2. How does the communications industry incorporate computer technology in its operation?
3. What are some of the factors that tend to delay the implementation of new technology?
4. What is videotex service? What equipment is necessary to access this service?
5. What are some of the videotex services used in the home?
6. What are some of the videotex services used by business and industry?
7. What is cellular radio and how does it differ from traditional mobile telephone service?
8. What are the chief advantages of cellular radio?
9. Describe voice message service. What are some of its advantages?
10. Name and briefly describe the three types of voice processing.
11. What is the role of telecommunications in the electronic office and management information systems?

References and Bibliography

Bylinsky, Gene. "Can Bell Labs Keep It Up?" *Fortune,* June 27, 1983, 90–91.

Cornish, Edward, ed. *Communications Tomorrow,* selections from *The Futurist.* Bethesda, Md.: World Future Society, 1982.

Didsbury, Howard F., Jr. *Communications and the Future.* Bethesda, Md.: World Future Society, 1982.

Edwards, Morris. "Communications Brings Integration to Office Automation Strategies." *Communications News,* April 1983, 48–54.

Ferrarini, Elizabeth M. "Getting in Touch with CompuServe." *Business Computer Systems,* October 1982, 35–36.

Frye, John. "The Impact of Telecommunications." *Technical Horizons in Education Journal,* Vol. 10, No. 6 (April 1983), 87–90.

Garner, Patricia A. "Telecommunications: A Vital Part of the Information Age." *The Secretary,* March 1984, 10–14.

Hindin, Harvey J. "Controlling the Electronic Office: PBXs Make Their Move." *Electronics,* April 7, 1981, 139–48.

International Data Corporation. "Business Communications: Challenges for the '80s." *Fortune,* April 4, 1983, 37–96.

Johnson, Margaret H., ed. *The Changing Office Environment,* NBEA Yearbook No. 18. Reston, Va.: National Business Education Association, 1980.

Kornbluh, Marvin. "The Electronic Office." *The Futurist,* June 1982, 37–42.

Martin, James T. *Future Developments in Telecommunications,* 2d ed. Englewood Cliffs, N.J.: Prentice-Hall, Inc., 1977.

———. *Telematic Society.* Englewood Cliffs, N.J.: Prentice-Hall, Inc., 1981.

Munsey, Wallace. "First Phase of Telemedicine Program is Prescription for Improved Health Care." *Communications News,* February 1983, 58B–58C.

Periphonics. *The Book on Voice Processing.* Bohemia, N.Y.: Periphonics, 1982.

Rosen, Arnold, Eileen Tunison, and Margaret Bahniuk. *Administrative Procedures for the Electronic Office.* New York: John Wiley & Sons, Inc., 1982.

Ross, Ian M. "Shaping the Information Society." *Bell Laboratories Record,* November 1982, 251–53.

Shaffer, Richard A. "Phones Seen as Major Market by Semiconductor Companies." *Wall Street Journal,* July 1, 1983, 11.

———. "Global Data System is Seen as Telephones Use More Digital Gear." *Wall Street Journal,* December 23, 1983, 1.

Singleton, Loy A. *Telecommunications in the Information Age.* Cambridge, Mass.: Ballinger Publishing Company, 1983.

Tunstall, Brooke. "The Shape of Things to Come." *Bell Telephone Magazine,* Vol 61, No. 3/4 (1982, No. 3/4), 14–19.

Williams, Frederick. *The Communications Revolution.* Beverly Hills, Calif.: Sage Publications, 1982.

Professional Associations

A
APPENDIX

The names of the current officers of most of the following organizations can be found in the *Encyclopedia of Associations,* published by Gale Research Co., Detroit, Michigan.

Associated Students for Career Orientation in Telecommunications (ASCOT)
Michigan State University
290 Communications Arts and Science Building
E. Lansing, MI 48424
(517) 355–8312

Association of College and University Telecommunication Administrators
37 N. Mills Street
Madison, WI 53706
(608) 262–0521

Association of Data Communications Users
P.O. Box 1184
New York, NY 10019
(201) 385–4540

Association of Long Distance Telephone Companies
2719 Soapstone Drive
Reston, VA 22091
(703) 476–5515

Computer and Communications Industry Association
1500 Wilson Blvd., Suite 512
Arlington, VA 22209
(703) 524–1360

Data Processing Management Association
505 Busse Highway
Park Ridge, IL 60068
(312) 825–8124

Information Industry Association
316 Pennsylvania Avenue, S.E., Suite 400
Washington, D.C. 20003
(202) 544–1969

Institute of Electrical and Electronics Engineers (IEEE)
Council on Communications
345 E. 47 Street
New York, NY 10017
(212) 705–7900

International Communications Association (ICA)
12750 Merit Drive
Suite 828, LB–89
Dallas, TX 75251
(214) 233–3889

International Organization of Women in Telecommunications (IOWIT)
2554 Lincoln Blvd., #30
Marina del Rey, CA 90292
(213) 306–0218

International Telecommunication Union (ITU)
Place des Nations
CH–1211 Geneva 20, Switzerland

Joint Council on Educational Telecommunications (JCET)
1126 16th Street, N.W.
Washington, D.C. 20036
(202) 659–9740

National Association of Radio and Telecommunications Engineers, Inc.
P.O. Box 15029
Salem, OR 97309
(503) 581–3336

National Association of Telecommunications Officers and Advisors (NATOA)
Office of Telecommunications
400 Yesler Building
Seattle, WA 98104
(206) 625–2268

National Communications Association (NCA)
485 Fifth Avenue, Suite 311
New York, NY 10017
(212) 682–2627

North American Telephone Association (NATA)
511 Second Street, N.E.
Washington, D.C. 20002
(202) 547–4450

Organization for the Protection and Advancement of Small Telephone Companies (OPASTCO)
1200 New Hampshire Avenue, N.W., Suite 320
Washington, D.C. 20036
(202) 659–5990

Society of Telecommunications Consultants (STC)
One Rockefeller Plaza, Suite 1912
New York, NY 10020
(212) 582–3909

Tele-Communications Association (TCA)
424 S. Pima Avenue
West Covina, CA 91790
(213) 919–2621

Telecommunications International Union (TIU)
2341 Whitney Avenue
Hamden, CT 06518
(203) 288-2445

United States Telephone Association (USTA)
1801 K Street, N.W., Suite 1201
Washington, D.C. 20006
(202) 872–1200

U.S. Telecommunications Suppliers Association (USTSA)
333 N. Michigan Avenue, Suite 1618
Chicago, IL 60601
(312) 782-8597

Organizations That Sponsor Seminars

B

APPENDIX

Advanced Training Professionals, Ltd., Inc.
111 East Avenue
Norwalk, CT 06851
(203) 866–6060

American Institute for Professional Education
100 Kings Road
Madison, NJ 07940
(201) 377–7400

AT&T Communications
Customer Education Center
15 West 6th Street
Cincinnati, OH 45202
(800) 543–0401; in Ohio (513) 352–7419

Architecture Technology Corporation
P.O. Box 24344
Minneapolis, MN 55424
(612) 935–2035

Business Communications Review
BCR Enterprises, Inc.
950 York Road
Hinsdale, IL 60521
(800) BCR-1234; In Illinois (312) 986–1432

CAPE (Center for Advanced Professional Education
11928 North Earlham
Orange, CA 92669
(714) 633–9280

Communications Solutions, Inc.
992 South Saratoga–Sunnyvale Road
San Jose, CA 95129
(408) 725–1568

Continuing Engineering Education
The George Washington University
School of Engineering and Applied Science
Washington, D.C. 20052
(800) 424–9773; locally 676–8516

Data Communication Seminars
445 West Main Street
Wyckoff, NJ 07481
(201) 891–8405

Datamation Institute
Seminar Coordination Office, Suite 415
850 Boylston Street
Chestnut Hill, MA 02167
(617) 738–5020

Datapro Seminars
Datapro Research Corporation
1805 Underwood Boulevard
Delran, NJ 08075
(800) 257–9406; in New Jersey (609) 746–0100

Friesen's School of Generic Telephony
Gerry Friesen
Building B, 1300 Chinquapin Road
Churchville, PA 18966
(215) 355–2886

GTE Telenet Communications
8229 Boone Boulevard
Vienna, VA 22180
(800) T-E-L-E-N-E-T; or (800) 835–3638

Institute for Advanced Technology
Control Data Corporation
6003 Executive Boulevard
Rockwood, MD 20852
(301) 468–8576

International Communication Association (ICA)
12750 Merit Drive
Suite 828, LB-89
Dallas, TX 75251
(214) 233–3889

Lee's abc of the Telephone
Training Administrator
P.O. Box 537
Geneva, IL 60134
(312) 879–9000

MCI School of Telecommunications Management
MCI Education Center
1301 Avenue of the Americas
New York, NY 10019
(212) 582–6520

Systems Technology Forum
9000 Fern Park Drive
Burke, VA 22015
(800) 336–7409; in Virginia (703) 425–9441

Telco Research Corporation
1207 17th Ave. S.
Nashville, TN 37212
(615) 329–0031

Telecom Library, Inc.
205 West 19 Street
New York, NY 10011
(800) LIBRARY; in New York (212) 691–8215

Tele-Strategies
6842 Elm St.
McLean, VA 22101
(703) 734–7050

Tellabs, Inc.
4951 Indiana Ave.
Lisle, IL 60532
(312) 969–8800, ext 241

United States Telephone Association (USTA)
1801 K Street, N.W.
Washington, D.C. 20006
(202) 872–1200

Telecommunication Equipment Suppliers

C
APPENDIX

Account-A-Call Corporation
4450 Lakeside Drive
Burbank, CA 91505

Anaconda-Ericson Information Systems
74655 Lampson Avenue
Garden Grove, CA 92641

AT&T Information Systems
1 South Parkway
Morristown, NJ 07960

ATC American Telecommunications, Inc.
9620 Flair Drive
El Monte, CA 91731

Atlantic Research Corporation
Teleproducts Division
5390 Cherokee Avenue
Alexandria, VA 22314

Avantek
481 Cottonwood Drive
Milpitas, CA 95035

Badger Meter, Inc.
Electronics Division
150 East Standard Avenue
Richmond, CA 94804

Centel Business Systems
O'Hare Plaza
5735 N East River Road
Chicago, IL 60631

Comdial Telephone Systems
1180 Seminole Trail
Charlottesville, VA 22906

CONTEL
Continentel Telecom, Inc.
245 Perimeter Center Parkway
Atlanta, GA 30346

Datapoint Corporation
9725 Datapoint Drive
San Antonio, TX 78284

Executone, Inc.
Two Jericho Plaza
Jericho, NY 11753

General DataCom Industries, Inc.
One Kennedy Avenue
Danbury, CT 06810

General Electric Corporation
One River Road
Schenectady, NY 12345

GTE Automatic Electric
400 N. Wolf Road
Northlake, IL 60164

GTE Business Communications Systems, Inc.
Retail Telephone Division
11601 Roosevelt Boulevard
North St Petersburgh, FL 33702

Harris Corporation
9541 Mason Avenue
Chatsworth, CA 91311

InteCom Incorporated
601 InteCom Drive
Allen, TX 75002

ITT Telecom
67 Broad Street
New York, NY 10004

Iwatsu America, Inc.
120 Commerce Road
Carlstadt, NJ 07072

Linatel Systems
4665 Nautilus Court
South Boulder, CO 80301

MAX
Melco Laboratories
14408 N.E. 20th Street
Bellevue, WA 98007

Micon Systems, Inc.
20151 Nordhoff Street
Chatsworth, CA 91311

Mitel
National Headquarters
600 West Service Road
Dulles International Airport
Washington, D.C. 20041

NEC
Nippon Electric Company, Ltd.
P.O. Box 1, Takanawa
Tokyo, Japan

NEC Telephones, Inc.
532 Broad Hollow Road
Milville, NY 11747

Northern Telcom
P.O. Box 10934
Chicago, IL 60610

North Supply Company
A United Telecom Company
600 Industrial Parkway, Industrial Airport
Kansas City, KA 66031

OKI
One University Place
Hackensack, NJ 07601

Redcom Laboratories, Inc.
750 Fairport Park
Fairport, NY 14450

ROLM Corporation
4900 Old Ironsides Drive
Santa Clara, CA 95050

SAN/BAR Corporation
2405 South Shiloh Road
Garland, TX 75401

Siemens Communications Systems, Inc.
1001 Broken Sound Parkway
Boca Raton, FL 33431

Solid State Systems, Inc.
"The Smart Telephone"
1990 Deck Industrial Boulevard
Marietta, GA 30067

Stromberg Carlson
A Plessey Telecommunications Company
400 Rinehart Road
Lake Mary, FL 32746

TeleSciences, Inc.
351 New Albany Road
Moorestown, NJ 08057

Telrad
Telecommunications and Electronic
Industries, Ltd.
P.O. Box 50
Lod, Israel

TIE/Communications, Inc.
5 Research Drive
Shelton, CT 06484

Tone Commander System, Inc.
4320 150th Avenue, N.E.
Redmond, WA 98052

Toshiba Telecom
2441 Michelle Drive
Tustin, CA 92680

Valcom, Inc.
1845 Production Street
Roanoke, VA 24013

D

APPENDIX

Sample Request for Proposal

This specification describes a telephone system to be installed at the XYZ Company, 4321 Main Street, Milltown, Michigan 48185.

1.1 *General Information.* The following describes the specifications and features to be included in a telephone system to be installed for the XYZ Company in Milltown, Michigan. A seven-year life is anticipated for this installation. These specifications are intended to be functional. Any deviation from the stated specifications and features that the vendor proposes must be fully explained in the proposal.

1.2 *Proposal Submission Date.* Proposals will be received at the XYZ Company until 12:00 noon Eastern Standard Time on ______, 198_. The selected vendor will be notified of intent to negotiate within one hundred twenty (120) days from this date; therefore, all prices submitted with proposals will be considered to be firm for that period.

1.3 *General System Description.* Proposals should be accompanied with a General System Description, along with specifications for the system offered by the vendor. (Sales literature is not acceptable unless it is complete and accurate in all details.)

2.0 *Systems Specifications.* The system shall be a stored program control type system utilizing solid state electronic components. The system should be designed to provide P.01 grade of service for the specified number of lines and telephones.

2.1 *Attendant Features.* The attendant console(s) shall be capable of the following: camp on; automatic recall; conference set up for 2 trunks and 3 internal telephones, or 1 trunk and 4 internal telephones; transfer of incoming and outgoing calls; alphanumeric display of originating telephone number; call-waiting queue; trunk group busy indicators; line load control keys; and busy-station verification.

2.2 *Switching System Features.* The switching system shall be capable of the following: automatic dialing between internal telephones; dial "9" capability for automatic connection to outgoing trunks; dial "0" capability for connection to attendant console; rotary service groups (hunting) for selection of internal telephones in a group in a sequential basis; station transfer without operator assistance; add-on capability without operator assistance; pushbutton dialing; automatic route selection (least-cost routing); and call detail recording capability.

3.0 *System Administration.* Changes such as station class of service, hunting sequence, telephone number, and other features to be made from a terminal located on the customer's premises.

3.1 *Traffic Measurement.* The system should be equipped with the necessary devices so that traffic counts can be made. The following information, by trunk groups, will be required:

a. Number of calls
b. Number of blocked calls
c. Number of times individual features are used
d. Number of calls handled at operator consoles, and
   (1) average delay on answered calls
   (2) number of abandoned calls
   (3) average waiting time before abandonment

4.0 *Telephone Sets.* All telephone sets to be dual-tone multifrequency (pushbutton) dialing. All key telephone sets to be equipped with lamps to flash 60 interruptions per minute when ringing, 120 on hold, and remain lighted when in use.

5.0 *Installation.* All installation work shall be performed in an orderly business manner and in accordance with applicable building codes. Location of the switch on the premises to be resolved with the selected vendor. All cabling to be concealed in conduit, building walls, ceilings, or similar areas. All feeder cable and distribution terminals to have a minimum of 30 percent spare capacity. Two (2) complete sets of "as installed" drawings, technical manuals, and spare part lists to be provided by vendor.

6.0 *Cutover Procedures.* Vendor will provide training materials for system users. Vendor will train XYZ key personnel who will conduct training sessions for other XYZ employees. Testing of system to be performed by the vendor to the satisfaction of the XYZ Company.

7.0 *Maintenance.* Proposals to specify maintenance provisions and terms of proposed maintenance contract. If vendor does not provide maintenance service, recommend alternate contractor. Specify terms of warranty. Provide diagnostic capability at site.

8.0 *Financial Arrangement.* Specify financial terms under rental, sell outright, lease direct, or lease through third party.

XYZ Company
Milltown, Michigan 48155
System Call Data

	at Cutover	*Ultimate*
Internal Calls		
Total Day	3,400	4,600
Busy Hour	550	700
Incoming Calls		
Busy Hour	3,000	4,000
Total Day	425	600
Outgoing Calls		
Busy Hour	3,600	5,000
Total Day	575	750
Total Calls		
Busy Hour	10,000	13,600
Total Day	1,550	2,050

System Equipment Requirements

	at Cutover	*Ultimate*
Station Lines	825	1,120
Telephones		
Single Line	700	920
6 Line	100	150
10 Line	25	50
Speakerphones	10	20
Attendant Consoles	2	3
Trunks		
Incoming (DID)	55	80
Outgoing		
Local	60	80
WATS/OCC	20	30
Tie Lines	4	6

Supplier's Questionnaire

This questionnaire is provided to assist in the evaluation of vendor responses to this Request for Proposal. Please list reference number in replying to each question. Use the following format in your answer:

1. Restate the question.
2. State answer, including cost sheets, if appropriate.
3. List reference.

1.1 Who is the manufacturer and what is the model number of the equipment to be supplied?

1.2 Does the system proposed comply with FCC registration requirements?

2.0 What features are available for the attendant consoles over those listed in the RFQ? List any features requested that cannot be provided.

3.0 What is the maximum traffic capacity per line at cutover, and what are the provisions to increase this capacity to meet growth to ultimate configuration? What is the maximum number of simultaneous conversations that can occur?

3.1 What types of calls can be recorded on the call detail recorder? What is the format and the capability of the system with respect to the preparation of the final cost allocation report? Please attach sample reports.

3.2 Can the least-cost routing feature handle calls through FX lines, OCC networks, WATS, and message toll? Does it provide time of day routing?

4.0 What will be the cable requirements for installation of the telephone sets? How much spare will be provided?

5.0 Please provide a diagram showing space requirements for the control unit. Specify power and environmental requirements.

6.0 What types of materials are available for training system users? What resources are available to train maintenance personnel?

7.0 What are the financial arrangements that you propose? In the event that a lease is requested, will you assign the lease to a third party? Please provide a proposed payment schedule.

APPENDIX

E Financial Analysis

Comparison of Monthly Rental, Purchase, or Lease of a Telephone System

Rental

Assumptions:

- ☐ A system that rents for $800 per month, including tax.
- ☐ Lines and trunks to rent for $1,000 per year.
- ☐ One-time installation charge of $2,500.
- ☐ Maintenance, insurance, and taxes included in rent.
- ☐ Income tax rate at 46 percent.
- ☐ Rental will escalate 6 percent per year due to inflation.

	1st yr.	2nd yr.	3rd yr.	4th yr.	5th yr.	6th yr.	7th yr.
Annual Rental	9,600	10,176	10,787	11,434	12,120	12,847	13,618
Rental Lines and Trunks	1,000	1,060	1,124	1,191	1,262	1,338	1,419
Installation	2,500						
Total Costs	13,100	11,236	11,911	12,625	13,382	14,185	15,037
Tax Credit (@ 46%)	6,026	5,169	5,479	5,808	6,156	6,525	6,917
Net Costs	7,074	6,067	6,432	6,817	7,226	7,660	8,120
Discount Factor (@ 15%)	1.000	.8691	.7561	.6575	.5718	.4972	.4323
Present Value	7,074	5,273	4,863	4,482	4,132	3,809	3,510

Total present value of all costs = $33,134

Purchase

Assumptions:

- ☐ A system that sells for $30,000, including tax.
- ☐ Lines and trunks to rent for $1,000 per year.
- ☐ Installation included in purchase price.
- ☐ Maintenance contract to cost $1,140 per year.
- ☐ Insurance to cost $100 per year.
- ☐ Property tax to be $1,200 per year
- ☐ Investment tax credit at 10 percent.
- ☐ Depreciation to be over 5 years on straight-line basis.
- ☐ Income tax rate at 46 percent.
- ☐ Maintenance, insurance, and taxes will escalate 6 percent per year due to inflation.

	1st yr.	2nd yr.	3rd yr.	4th yr.	5th yr.	6th yr.	7th yr.
Purchase Price	30,000						
Rental Lines and Trunks	1,000	1,060	1,124	1,191	1,262	1,338	1,419
Maintenance		1,208	1,281	1,358	1,439	1,526	1,617
Insurance	100	106	112	119	126	134	142
Property Taxes	1,200	1,272	1,348	1,429	1,515	1,607	1,702
Total Expense	2,300	3,646	3,865	4,097	4,342	4,605	4,880
Total Costs	32,300	3,646	3,865	4,097	4,342	4,605	4,880
Investment Tax Credit	3,000						
Depreciation	6,000	6,000	6,000	6,000	6,000		
Tax Credit (@ 46%)	5,198	4,437	4,538	4,645	4,757	2,118	2,245
Net Costs after tax	27,102	(791)	(673)	(548)	(515)	2,487	2,635
Discount Factor (@ 15%)	1,000	.8691	.7561	.6575	.5718	.4972	.4323
Present Value	27,102	(687)	(509)	(360)	(294)	1,237	1,139

Total present value of all costs = $27,628

Lease

Assumptions:

- ☐ A system that sells for $30,000, including tax.
- ☐ Lines and trunks to rent for $1,000 per year.
- ☐ Installation included in purchase price.
- ☐ Maintenance contract to cost $1,140 per year.
- ☐ Purchase option at end of five years at 10% of purchase price.
- ☐ Insurance and taxes to be assumed after title passes.
- ☐ Income tax rate at 46 percent.

	1st yr.	2nd yr.	3rd yr.	4th yr.	5th yr.	6th yr.	7th yr.
Purchase Price						3,000	
Rental Lines and Trunks	1,000	1,060	1,124	1,191	1,262	1,338	1,419
Annual Lease	9,600	9,600	9,600	9,600	9,600		
Maintenance		1,208	1,281	1,358	1,439	1,526	1,617
Insurance						134	142
Property Taxes						1,607	1,702
Total Expense	10,600	11,868	12,005	12,149	12,301	4,605	4,880
Total Cost	10,600	11,868	12,005	12,149	12,301	7,605	4,880
Tax Credit (@ 46%)	4,876	5,459	5,522	5,589	5,658	2,118	2,245
Net Costs after tax	5,724	6,409	6,483	6,560	6,643	5,487	2,635
Discount Factor (@ 15%)	1,000	.8691	.7561	.6575	.5718	.4972	.4323
Present Value	5,724	5,570	4,902	4,313	3,798	2,728	1,139

Total present value of all costs = $28,174

Glossary

Note: An asterisk after a definition indicates that it is an American National Standards Institute definition. The number listed in parentheses after the definition is the page number on which the term first appears in this text.

A

Access code The preliminary digits that a user must dial to be connected to a particular outgoing trunk or line.* (72)

Alternating current (AC) Electrical current that travels first in one direction (+ to −) and then in the other direction (− to +). (106)

Amplifier See *repeater.* (103)

Amplitude The maximum variation from the zero position of any alternating current. The size or magnitude of an alternating wave form. (106)

Amplitude modulation A form of modulation in which the amplitude of a carrier wave is varied in accordance with some characteristic of the modulating signal.* (109)

Analog signal A continuous electrical signal that varies in voltage to reflect variations in some quantity such as loudness of the human voice. (69)

ASCII American Standard Code for Information Interchange. (Pronounced "ask-ee.") An 8-bit code (one bit is for parity check) developed by the American Standards Association that has been adopted as the standard code for data transmission in the United States. (154)

Asynchronous transmission A transmission process in which there is a variable time interval between successive bits. It is often known as start-stop transmission. (152)

Attenuation The difference between transmitted and received power due to transmission loss through communications equipment. (103)

Audio frequencies Frequencies that can be heard by the human ear, about 20 to 20,000 Hz. (107)

Automatic Identification of Outward Dialing (AIOD) A PBX service feature that identifies the calling extension, thereby permitting the cost of the call to be allocated to the extension. (173)

Automatic route selection A PBX service feature that permits automatic selection of the most efficient routing of a call in a corporate network. It is sometimes called least-cost routing. (76)

Average busy-hour traffic count The average number of calls received during the busy hour over a specified number of days. (227)

B

Bandwidth The range between the lowest and highest frequencies of a channel. (106)

Baseline The starting point for a sine wave. (155)

Batch processing A method of processing data in which input records are collected in their original, physical form over a period of time, transcribed onto an input medium that the computer can read, and then transported to the computer room in groups that are entered into the computer for processing. (120)

Baud A unit of signaling speed derived from the reciprocal time of the shortest pulse width in the bit stream. (147)

Baudot code A 5-bit, 32-character alphanumeric code used in asynchronous teleprinter transmission. (153)

Beeper A slang term for a radio pager that signals the wearer that he or she has a message waiting. (274)

Bit A contraction of BInary digiT. The smallest unit of information in a code using the binary system. It represents one of two possible values such as a mark or a space, a 1 or a 0, or an on or an off. (70)

Block A group of continuous characters transmitted as a unit. (151)

Block character checking A method of error detection in data transmission based on the observance of preset rules for formation of blocks.* (151)

Blocked call A call that cannot be completed because of a network busy condition. (209)

Blocking The inability to complete a connection between two lines because of a network busy condition. (65)

Box telephone Telephone instrument developed and patented by Alexander Graham Bell. (29)

Broadband A synonym for wideband. A communication channel having a bandwidth broader than that of a voice-grade channel, thereby providing high-speed data transmission capability. (107)

Buffer A temporary storage device used to compensate for a difference in rate of flow of data or time of occurrence of events when transmitting data from one device to another. (137)

Busy hour The two consecutive half-hour periods of a day in which the largest number of calls occur. (199)

C

Call capacity The ability of a telephone system to handle a specific number of calls to provide a specific grade of service. (214)

Call forwarding A telephone service feature that permits automatic forwarding of calls to another telephone number. (174)

Calling rate The number of calls per telephone; determined by dividing the count of busy-hour calls by the number of telephones. (209)

Call waiting (camp-on) A telephone service that permits a call to a busy telephone to be held while an audible tone notifies the busy telephone that a call is waiting. (75)

CCITT Consultative Committee on International Telegraphy and Telephone. A part of the International Telecommunications Union (ITU), which is an organ of the United Nations. CCITT is the forum for recommendations for international communication systems. (72)

CCS Abbreviation for hundred-call seconds. (230)

Cell A subdivision of a mobile telephone service area; it contains a low-powered radio communicating system connected to the local telephone network. (272)

Cellular radio An advanced form of mobile telephone service combining radio and computer technology to provide telephone service to moving vehicles. (271)

Central office A synonym for switching center, also referred to as a telephone exchange. (63)

Central office switching equipment The mechanical, electromechanical, or electronic equipment that routes a call to its ultimate destination. (65)

Centralized processing A data processing configuration wherein the processing for several divisions, functional units, or departments is centralized on a single computer, with input/output devices located in the same area as the computer. See also *distributed processing.* (118)

Central processing unit (CPU) The component of a computer that does the actual processing of data. (122)

Centrex A unit of telephone equipment that permits subscribers to be directly dialed from the outside. (74)

Channel A single unidirectional or bidirectional path for transmitting or receiving or both of electrical signals. (101)

Character checking A method of error detection in data transmission using preset rules for checking of characters. (150)

Check bit One noninformation-carrying bit added to characters being transmitted that enables the computer to run its own check on every character it processes. Also called a parity bit. (150)

Circuit The complete path between two end-terminals over which one-way or two-way communication may be provided. (101)

Class of service restriction A feature that limits use of a telephone station to certain types of calls. (174)

Coaxial cable A cable consisting of one or more hollow cylinders with a single wire running down the center of each cylinder. It can carry a much higher bandwidth than a wire pair. (91)

Code Any system of communication in which arbitrary groups of symbols represent units of plain text of various length.* (147)

Code set The complete set of representations defined by a code.* (147)

Coding The process of converting information into a form suitable for communications.* (147)

Common-battery A DC power source in the central office that supplies power to the central office switching equipment and to all subscribers connected thereto. (63)

Common carrier An organization in the business of providing communications services to the public and which is regulated by the appropriate state or federal agency. (45)

Common carrier principle The regulatory concept that limits the number of companies that can provide certain essential services in a specific geographic area. (46)

Common Control Switching Arrangement (CCSA) A private network provided by the telephone company that shares switching facilities with the public network. (89)

Communication satellite An orbiting vehicle that relays signals between communications stations. (95)

Communications control unit (CCU) See *front-end processor.*

Computer conferencing A visual form of conference telephone call, in which conferees talk to each other by keyboarding messages and transmitting them over telecommunications facilities to other computers connected to the system. (270)

Computerized Branch Exchange (CBX) A PBX that uses a small computer with a solid-state switching network. (74)

Conditioned line A private line specially treated to reduce distortion and improve transmission quality. (89)

Control character A character used to define a subsequent series of characters until the next control character appears. (153)

Control unit A component of modern telephone instruments that allows callers to place calls directly, without the assistance of an operator. It can be either a rotary dial or pushbutton keys. (63)

Country code The second set of digits a customer dials to place an international call (following the international access code). (73)

Creamskimming The practice of competing only in the most profitable markets. (252)

Crossbar system A type of common-control switching system using switches that have a plurality of vertical paths and a plurality of horizontal paths interconnected to form a communications link. (68)

Crosstalk The phenomenon in which a signal transmitted on one circuit or channel of a transmission system creates an undesired effect in another circuit or channel.* (89)

Cursor A blinking symbol that indicates current location on the CRT screen. (137)

Customer-premises equipment (CPE) Terminal equipment installed on the customer's premises that is connected to the telephone network. It may be obtained from any supplier. (73)

Cutover The activation of a newly installed telephone system that either replaces an older system or is a new installation. (222)

D

Data Any representations, such as characters or analog quantities, to which meaning is or might be assigned.* (116)

Data communications The movement of encoded information by means of electrical transmission systems. (116)

Data set A Bell System synonym for modem. (70)

Dedicated circuit or line A telecommunications channel used exclusively by a single subscriber. (86)

Delivery time The time from the start of transmission at the transmitting terminal to the completion of reception at the receiving terminal, when data is flowing in only one direction. (120)

Demodulation A function of changing the band pattern of a message on a carrier wave back into the form of the original message signal after transmission. (109)

Dial-tone first service A coin telephone service that permits customers to reach the operator and to dial certain calls, such as directory assistance or 911, without depositing a coin. (168)

Dibits A method used by some transmission systems to combine groups of bits and transmit two bits at a time, thereby doubling the transmission speed. (147)

Digital signal A nominally discontinuous electrical signal that changes from one state to another in discrete steps.* (70)

Direct current (DC) Electrical current that travels in only one direction in a circuit (+ to −). (106)

Distributed processing The processing of data at remotely located sites using communication lines to interconnect microcomputers or intelligent terminals with the central computer. See also *centralized processing.* (6)

E

Earth station A satellite antenna, consisting of a large dish that points at the satellite. (97)

EBCDIC An acronym for Extended Binary Coded Decimal Interchange Code. An 8-bit data transmission code used in IBM systems. (154)

Echo A type of distortion; an electric wave that has been reflected back to the transmitter with sufficient magnitude and delay to be perceived. (104)

Echo canceler Device that performs the same function as an echo suppressor, but unlike that device does not clip speech of the speaker and can work during two-way transmissions. (104)

Echo suppressor A device installed by telephone companies to reduce echo to a negligible level. (104)

Echo suppressor disabler A device that transmits a tone that can be heard on the telephone as a high-pitched whistle. The tone disables the echo suppressor until there has been no signal on the line for about 50 milliseconds. (104)

Electronic mail The delivery of mail, at least in part, by electronic means. (175)

Electronic switching system (ESS) Any switching system whose major components utilize semiconductor devices.* (264)

Equivalent Queue Extended Erlang B Tables developed by Dr. James Jewett to be used in the design of trunks that automatically route blocked calls to alternate routes. (238)

Erlang, Erlang B, and Erlang C tables Tables used to predict quantities of equipment required to produce a desired grade of service at a given level of traffic. (230)

Exchange area A geographical area that has a single uniform set of charges for telephone service. (163)

Extended Area Service (EAS) A service that permits a subscriber to make calls to a designated area beyond the local exchange area and be charged local exchange rates instead of toll rates. (165)

F

Facility A transmission path between two or more locations. (101)

Facsimile (FAX) The process of transmitting text, pictures, diagrams, etc., via a telecommunication system to a remote location where a hard copy of the transmitted material is reproduced. (4)

Federal Communications Commission (FCC) A board of five commissioners appointed by the president of the United States under the Communications Act of 1934, having the power to regulate interstate and foreign electrical communication systems originating in the United States. (10)

Foreign exchange service (FX) A service providing a circuit connecting a subscriber's main station for private branch exchange with a central office of an exchange other than that which normally serves the exchange area in which the subscriber is located. (88)

Fiber optics Hair-thin filaments of transparent glass or plastic that use light instead of electricity to transmit voice, video, or data signals. (99)

Flat rate service Service wherein the user is entitled to an unlimited number of telephone calls within a specified local service area for a fixed monthly rate. (164)

Foreign attachment Equipment not provided by the telephone company that is attached to telephone lines. (54)

Frequency The number of cycles or events per unit of time. When the unit of time is one second, the measurement unit is the hertz (Hz).* (106)

Frequency-division multiplexing One of the two basic multiplexing techniques, in which the channel frequency range is divided into narrower frequency bands. See also *time-division multiplexing.* (110)

Frequency modulation (FM) A process in which the intelligence of a signal is represented by variations in the frequency of the oscillation of the signal. (109)

Front-end processor A programmed-logic or stored-program device that interfaces data communication equipment with an input/output device or memory of a data processing computer.* (145)

Full-duplex (FDX) A type of operation in which simultaneous two-way conversations, messages, or information may be passed between any two given points.* (104)

G

Grade of service The probability of a call being blocked or delayed, expressed as a percentage. (200)

H

Half-duplex (HDX) A circuit that affords communication in either direction but only in one direction at a time.* (104)

Handshaking An exchange of predetermined characters or signals between two stations to provide control or synchronism after a connection is established.* (149)

Harmonic telegraphy The transmission of a number of telegraph messages over a single wire simultaneously using interrupted tones of different frequencies. (27)

Hertz (Hz) A unit of measurement formerly called cycles-per-second. (69)

Holding time Telephone conversation time plus the time that equipment is used to establish connection. (228)

Hookswitch See *switchhook.*

I

Identified ringing A telephone service feature that provides distinctive ringing tones for different categories of calls. (172)

Independent telephone company Prior to AT&T's divestiture, a company providing common carrier telephone service to subscribers independent of any Bell affiliation and in direct competition with Bell. (36)

Individual service One telephone line to serve one subscriber. (165)

Information Processed data (as opposed to raw data). (116)

Information utility A commercial firm that provides time-sharing services on a computer. (129)

Inhouse system A computer within an organization that is usually time-shared among several departments. (129)

Intelligent terminal A terminal that contains a processing unit and can perform data processing and storage functions. (140)

Interactive system A realtime telecommunication system that provides immediate, two-way communication between terminals and a computer, processing transactions as they occur. (122)

Interconnect company A company that provides telecommunications terminal equipment for connection to phone company lines. (55)

International access code The first set of digits a customer dials to place an international call. (73)

International Telecommunication Union (ITU) The telecommunications agency of the United Nations established to promote standardized telecommunications on a worldwide basis. (72)

K

Key telephone set A telephone set with buttons or keys located on or near the telephone. It is used with associated equipment to provide features such as call holding, multiline pickup, signaling, intercommunication (intercom), and conferencing. (76)

Kingsbury Commitment A 1913 letter from AT&T vice-president Nathan C. Kingsbury to the U.S. attorney general in response to accusations that AT&T was a monopoly. Kingsbury committed AT&T to disposing of its stock in Western Union, refraining from acquiring independent telephone companies without approval of the ICC, and interconnecting its facilities with those of the independents. (41)

L

Large scale integration (LSI) The integration of thousands of circuits onto a single chip. (258)

Laser The acronym for Light Amplification by Stimulated Emission of Radiation. Used to generate very high frequency beams of light with tremendous information capacity. (100)

Leased line See *private line.* (86)

Least-cost routing See *automatic route selection.* (76)

Line discipline The sequence of operations involving the actual transmitting and receiving of data. Sometimes synonymous with protocol. (149)

Link The communication facilities existing between adjacent nodes.* (85)

Load/service relationship analysis Test that a telecommunications manager conducts to determine whether a telephone system has been engineered to provide the desired grade of service. (240)

Local Access and Transport Areas (LATAs) Local calling areas mapped into 160 local access and transport areas throughout the United States. Only long distance companies are empowered to provide telephone service between LATAs. (253)

Local area network (LAN) A configuration of telecommunications facilities designed to provide internal communications within a limited geographical area. (89)

Local exchange service Public telephone service to points within the designated local service area (exchange area) for a telephone. (163)

Local messages The total charges on a telephone bill for the calls made within the local calling area. (193)

Local service area A geographical area that has a single, uniform set of charges for telephone service. (163)

Long distance access code A code used to gain access to a long distance network of a specific common carrier. (72)

Longitudinal redundancy checking (LRC) See *block checking.* (151)

Loop A channel between the customer's terminal and the central office. (63)

M

Magneto A component of early telephones that was a hand-operated electrical generator. Callers cranked the handle on the generator to activate a bell or light that signaled an operator. (63)

Mainframe computer Large computers that are capable of processing large amounts of data with very fast processing speeds, but that require a special environment and staff with data processing skills. See also *microcomputer* and *minicomputer.* (118)

Management information systems (MIS) The formal management of the flow of information throughout an organization, usually coordinated by a management information system department. (6)

Measured local service Telephone service for which a charge is made in accordance with a measured amount of usage, referred to as message units. (165)

Measured rate service Telephone service wherein a charge is made in accordance with a measured amount of usage. (164)

Medium A component of data communication systems, the path over which information flows. (133)

Message box In a voice message service, the designated area in a system's computer that receives a spoken message. (272)

Message switching A method of handling message traffic through a switching center, either from local users or from other switching centers, whereby a connection is established between the calling and called stations or the message traffic is stored and forwarded through the system.* (117)

Message systems Electrical transmission systems, such as TWX and Telex, that send messages, as opposed to conversation, in data form. (4)

Message unit A unit of measurement used in charging for local telephone calls. Criteria used are length of the call and the distance involved. (164)

Microcomputer The smallest general-purpose computer. Often it serves as a special-purpose or single-function computer on a single chip. See also *minicomputer* and *mainframe computer.* (118)

Microprocessor An electronic device consisting of a central processing unit, memory circuits, and input-output devices. (259)

Minicomputer A medium-sized class of computers that are larger and more expensive than microcomputers but smaller and less expensive than mainframes. See also *mainframe computer* and *microcomputer.* (118)

Mobile telephone service Telephone service between stationary telephones and moving vehicles that uses both the telephone network and a radio circuit to establish communication. (80)

Modem Acronym for MOdulator/DEModulator. A device that modulates and demodulates signals. They are primarily used for converting digital signals into analog signals for transmission and reconverting the analog signals into digital signals.* (70)

Modulation The process, or results of the process, of varying certain characteristics of a signal, called a carrier, in accordance with a message signal.* (108)

Multidrop circuit or line See *multipoint circuit or line.*

Multiplexer A device that combines a number of low-speed channels into one higher speed channel at one end of a transmission and divides it back into low-speed channels at the other. (110)

Multiplexing Use of a common channel to make two or more channels, either by splitting of the frequency band transmitted by the common channel into narrower bands, each of which is used to constitute a distinct channel, or by allotting this common channel to multiple users in turn, to constitute different intermittent channels.* (110)

Multipoint circuit or line A circuit providing simultaneous transmission among three or more separate points.* (157)

Multipoint network A line of a network with more than one terminal on it that is connected to the computer system. (157)

Multiprocessing A process wherein two or more CPUs are interconnected into a single system and one control program operates both processors. (123)

Multiprogramming The execution of two or more programs on the same computer simultaneously by interleaving their operations. (122)

N

Narrowband A channel whose bandwidth is less than that of a voice-grade channel. (107)

National Telecommunications and Information Administration (NTIA) An organization formed in March 1978, combining the Office of Telecommunications Policy and the Office of Telecommunications of the Commerce Department to provide advisory assistance in telecommunications and information issues for the Department of Commerce. (52)

Network A series of points connected by communications channels. (117)

Node In network topology, a terminal of any branch of a network or a terminal common to two or more branches of a network.* (85)

Noninteractive system A system (such as an offline system) where no interaction takes place between the user at a terminal and the computer during execution of a program. (120)

Numbering plan area (NPA) A geographic division within which telephone directory numbers are subgrouped. A three-digit code is assigned to each numbering plan area. (71)

O

Office of Telecommunications Policy (OTP) A federal organization formed by President Nixon in 1970 to assist the government in formulating policies for the telecommunications industry. (51)

Offline That condition wherein devices or subsystems are not connected into, do not form a part of, and are not subject to the same controls as an operational system. These devices may, however, be operated independently.* (120)

Online That condition wherein devices or subsystems are connected into, form a part of, and are subject to the same controls as an operational system.* (121)

Optical fiber See *fiber optics.*

Originating restriction A feature that restricts the telephone station from being used to place outgoing telephone calls. (174)

Other Common Carriers (OCCs) Common carriers other than Bell System carriers. (55)

Overflow A call that cannot be completed because of a network busy condition. (209)

Overnight telegram A telegraph service with following morning delivery. (178)

Overrun An expense that exceeds a budgeted amount. (201)

P

Panel system An early type of electromechanical switching equipment in which groups of numbers were arranged on frames resembling panels. (67)

Parity In binary-coded systems, a condition obtained with a self-checking code such that in any permissible code expression the total number of 1's or 0's is always even or always odd.* (150)

Parity bit A single bit used to detect errors in data transmissions. (150)

Parity checking A classical method of error detection during data transmission. (150)

PBX Private branch exchange. A small telephone exchange installed on a customer's premises to allow internal dialing from station to station and connection to incoming and outgoing lines. (73)

Phase The relative timing of an alternating signal. (155)

Phase modulation (PM) A form of modulation in which the phase or timing of the signal is shifted to respond to the pattern of the intelligence being transmitted. (109)

Point to point Simplest type of network, in which lines directly connect two points in a data communication network. (157)

Poisson tables Tables used to predict quantities of equipment required to produce a desired grade of service at a given level of traffic. (234)

Polling Calling up terminals in sequence to request the terminal to transmit a message. Usually, a central control unit polls terminals. (136)

POTS Plain Old Telephone Service. A term used to describe the basic service of supplying a single telephone set and access to the public-switched network. (4)

Private automatic branch exchange (PABX) An automatic PBX. (74)

Private branch exchange See *PBX*. (73)

Private line See *dedicated circuit*. Do not confuse private line with a one-party line. (86)

Private network A configuration of private lines and related switching facilities that are provided for the exclusive use of one customer. (167)

Protocol The rules for communication system operation that must be followed if communication is to be effected.* (149)

Public coin telephones Telephones that provide service on the public network to persons away from their residence or place of business. (168)

Public network The traffic network that provides public telephone service. Also called public-switched network. (86)

Public service commission (PSC) An agency charged with regulating communications services, as well as other public utility services, usually within a state. (52)

Public utility commission Same as public service commission. (52)

Pulse code modulation That form of modulation in which the modulating signal is sampled, the sample quantified and coded, so that each element of information consists of different kinds of numbers of pulses and spaces.* (70)

Q

Queuing The process of holding telephone calls in a waiting line until they can be answered. (236)

R

Rate center A geographically specified point used for determining mileage-dependent telephone rates. (249)

Rate of return The ratio of net profit to the total invested capital. The maximum rate of return for telecommunications companies is specified by the appropriate regulatory agency. (246)

Raw data Unprocessed data (as opposed to information, which is processed data). (116)

Realtime processing Processing that occurs at the same time that a transaction is taking place. (121)

Remote access Pertaining to communcation with a data processing facility through a data link.* (172)

Remote job entry (RJE) In computer operations, that mode of operation that allows input to a computer of a job from a remote site and receipt of the output at a remote site via a communcations link.* (120)

Remote terminal Device with a typewriter-like keyboard used for entering data some distance from the mainframe CPU where it will be processed. (134)

Repeater A device that amplifies an input signal or, in the case of pulses, amplifies, reshapes, retimes, or performs a combination of any of these functions on an input signal for retransmission.* (103)

Response time In a data system, the elapsed time between the end of transmission of an inquiry message and the beginning of the receipt of the response message, measured at the inquiry originating station.* (120)

Rotary dial A type of calling device that, when wound up and released, generates pulses required for establishing connection in a telephone system. (63)

Routing code The area code that comprises the third group of digits a customer dials to place an international call, or the first set of digits a customer dials for a long distance call within the same nation. (73)

S

Satellite An object or vehicle orbiting, or intending to orbit, the earth, moon, or other celestial body.* (95)

Satellite earth terminal The portion of a satellite link that receives, processes, and transmits communications between the earth and a satellite. (97)

Satellite relay An active or passive satellite repeater that relays signals between two earth terminals.* (95)

Semipublic coin telephones Telephones installed where there is a combination of general public and individual customer need for the service; the subscriber receives a listing in the telephone directory and guarantees a specific monthly revenue from the telephone. (168)

Service bureau A commercial firm that provides time-sharing services on a computer. (129)

Simplex A circuit using ground return and affording communications in either direction, but in only one direction at a time.* (104)

Sine wave An undulating wave used to represent the frequency of oscillation of an alternating current. (106)

Sink A component of data communication systems, the receiver of the information. (133)

Soft copy A visual display on a CRT screen that provides no permanent record of the information that is displayed. (136)

Source A component of data communication systems, the originator of the information. (133)

Specialized Common Carrier (SCC) A common carrier that specializes in a specific type of telecommunications service such as long distance. (12)

Speed calling A telephone service feature that permits a caller to reach certain frequently called numbers by using abbreviated telephone codes in place of the conventional telephone number. (171)

Station One of the input or output points on a communications system. (85)

Station Message Detail Recording (SMDR) Same feature as Automatic Identification of Outward Dialed Calls (AIOD). (173)

Step-by-Step An automatic switching system in which a call is extended progressively step-by-step to the desired terminal under direct control of pulses from a customer's dial. (67)

Store-and-forward A communciation service in which messages are received at intermediate points and stored for later retransmission to a further point or to their ultimate destination. (117)

Stored program control Electronic switching equipment that can be programmed to perform a variety of functions in addition to conventional call completion. (69)

Strowger switch A step-by-step switch named after its inventor, Almon B. Strowger. (67)

Switched network See *public network.*

Switchhook A switch on a telephone set that signals the central office that the telephone is either idle or in use. It is operated by the removal or replacement of the receiver or handset on the support mechanism. (Sometimes referred to as *Hook-switch.)* (63)

Switching The process of transferring a connection from one telephone circuit to another by interconnecting the two circuits. (65)

Switching center An installation in which switching equipment is used to interconnect communication circuits on a message or circuit switching basis.* (65)

Synchronization The process of determining and maintaining the correct timing for transmitting and receiving information. (152)

Synchronous transmission A transmission process that ensures that between any two significant instants in the overall bit stream, there is always an integral number of unit intervals.* (152)

T

Talking paths A network of interconnected paths forming a communications link in a switching system. (68)

Tariffs The published rates, regulations, and descriptions governing the provision of communications services. (53)

Telecommunication Any transmission, emission, or reception of signs, signals, writings, images, and sounds or information of any nature by wire, radio, visual, or the electromagnetic system.* (3)

Telecommuting Installation of a computer system in the home that allows an employee to communicate with the office without actually traveling to work. (270)

Teleconferencing A conference between persons remote from one another but linked by a telecommunications system.* (79)

Telecopier Facsimile machine. (175)

Telecourses Instruction delivered to the home via telecommunications. (268)

Telegraph A system of communication using coded signals.* (22)

Telematics The marriage of telecommunications and computer technology. The word is derived from *telematique,* a French term that describes the merging of telecommunications with computers and television. (5)

Telemedicine Provision of health care from a distance, linking a remote location by two-way television and audio signals with a hospital or health-care faciltiy. (267)

Telephone exchange A room or building equipped so that telephone lines terminating there may be interconnected as required.* (65)

Telephony The science and practice of transmitting speech or other sounds over relatively large distances, i.e., distances normally greater than earshot range, and rendering the sounds audible upon receipt.* (4)

Teleprinter See *teletypewriter.* (115)

Teleprocessing The overall function of an information transmission system that combines telecommunications, automatic data processing, and human-machine interface equipment and their interaction as an integrated whole.* (117)

Teletext A one-way transmission system in which data signals are transmitted over the FM portion of a television signal. (266)

Teletraffic theory The mathematical description of message flow in a communication network. (226)

Teletypewriter A printing telegraph instrument having a signal-actuated mechansim for automatically printing received messages. It may have a keyboard similar to that of a typewriter for sending messages. (The term "teleprinter" may be applied to a receive-only unit having no keyboard.)* (115)

Terminal equipment Communications equipment at each end of a circuit to permit the stations involved to accomplish the mission for which the circuit was established.* (Sometimes called "terminal.") (118)

Tie line Same as dedicated line and full-time circuit. (86)

Time-division multiplexing One of the two basic multiplexing techniques, in which the channel frequency is assigned successively to several different users at different times. See also *frequency-division multiplexing.* (11)

Traffic The flow of messages through a communication system. (226)

Traffic engineering The science of designing facilities to meet user requirements. (227)

Transponder A combination receiver-transmitter that receives a signal, amplifies it, and retransmits it at a different frequency. Communication satellites serve as transponders. (95)

Trunk A single transmission channel between two points, both of which are switching centers or nodes, or both.* (64)

U

Unbundling Separation of services by pricing rates for switching services separately from prices for customer-premises equipment. (253)

V

Value-added carrier A specialized common carrier that provides a service over and above the transmission of voice or data. The added value is usually computer-oriented. (50)

Vertical services Services over and above what is required for basic communications capability; e.g., deluxe telephone station sets or custom calling services. (250)

Videoconferencing A form of teleconferencing using video facilities. (79)

Videotex A generic term used to describe a group of consumer-oriented electronic information and transaction services. (266)

Viewdata A form of videotex service offered by the British Post Office that uses telephone lines for transmission. (266)

Voiceband A channel with a bandwidth appropriate for audio transmission, generally with a frequency range of about 300 to 3,000 Hz. (106)

Voice message service An advanced form of telephone answering service that permits a caller to send a one-way spoken message to a service user; the message is stored in the designated message box. (272)

Voice recognition system A telephone service using speech recognition to activate equipment that dials telephone numbers automatically. (272)

Voice response The conversion of computer output into spoken words and phrases that a human being can hear and understand; it is a combination of various frequencies of electrical impulses. (273)

Voice store-and-forward systems A system that enables a computer to accept a message and store it until a transmission path is available or the desired party calls to retrieve it. (273)

W

Waveguide A transmission line consisting of a hollow metallic conductor, generally rectangular, elliptical, or circular in shape, within which electromagnetic waves may be propagated.* (98)

Wide Area Telecommunications Service (WATS) A service that permits customers to make (OUTWATS) or receive (800) long distance voice or data calls and to have them billed on a bulk rather than individual call basis. The service is provided within selected service areas, or bands, by means of special private-access lines connected to the public telephone network. A single-access line permits inward or outward service, but not both. (76)

Wideband A synonym for broadband. Any analog signal or analog representation of a digital signal whose essential spectral content is broader than what can be contained within a voice channel of 4 kHz.* (107)

Workstation A terminal situated for use by different staff with different responsibilities, usually to access a computer data base. (6)

Index

A

B

C

G

H

I

K

L

M

N

O

P

R

S

T

U

V

W